W. Madill

Applied mechanics
Level 3

Longman London and New York

Longman Group Limited
Longman House, Burnt Mill, Harlow
Essex CM20 2JE, England
Associated companies throughout the world

*Published in the United States of America
by Longman Inc., New York*

© Longman Group Limited 1984

All rights reserved. No part of this publication
may be reproduced, stored in a retrieval system, or transmitted
in any form or by any means, electronic, mechanical,
photocopying, recording, or otherwise, without the
prior permission of the Copyright owner.

First published 1984

British Library Cataloguing in Publication Data

Madill, William
   Applied mechanics, level 3.
   1. Mechanics, Applied
   I. title
   620.1   TA350

ISBN 0-582-41292-7

Set in Compugraphic Times
Printed in Hong Kong
by Wing Lee Printing Co Ltd

To Valerie, James and Peter

General Editor—Construction and Civil Engineering
C.R. Bassett, B.Sc., F.C.I.O.B.
Formerly Principal Lecturer in the Department of Building and Surveying, Guildford County College of Technology

Books published in this sector of the series:

Building organisations and procedures   *G. Forster*
Construction site studies – production, administration and personnel   *G. Forster*
Practical construction science   *B.J. Smith*
Construction surveying   *G.A. Scott*
Materials and structures   *R. Whitlow*
Construction technology Volumes 1, 2, 3, and 4   *R. Chudley*
Maintenance and adaptation of buildings   *R. Chudley*
Building services and equipment Volumes 1, 2, and 3   *F. Hall*
Measurement Level 2   *M. Gardner*
Measurement Level 3   *M. Gardner*
Structural analysis Level 4   *G.B. Vine*
Economics for the construction industry   *R.C. Shutt*
Design procedures Level 4   *J.M. Zunde*
Design technology Level 5   *J.M. Zunde*
Architectural design procedures   *C.M.H. Barritt*
Construction technology for civil engineering technicians   *P.L. Monckton*
Environmental science   *B.J. Smith, G.M. Phillips and M. Sweeney*
Concrete technology Level 4   *J.G. Gunning*
Construction drawing Level 1   *R. Boxall*
Construction processes Level 1   *R. Greeno*
Design of structural elements 1   *A.G. Smyrell*
Environmental science   *B.J. Smith, G.M. Phillips and M. Sweeney*
Basic accounting for builders   *D. Hughes*
Building technology Level 4   *C.M.H. Barritt*
Site surveying Level 3   *P. Neal*
Site surveying and levelling Level 2   *H. Rawlinson*
Structural analysis Level 5   *S.R. Mangalgiri*

Contents

Preface x

Part 1 Kinematics

Chapter 1 Velocity and acceleration
Learning objectives 3
1.1 Scalars and vectors 3
1.2 Linear displacement and distance 4
1.3 Linear velocity and speed 4
1.4 Linear acceleration 4
1.5 Distance – time curve 4
1.6 Speed – time curve 5
1.7 Equations for linear, uniformly accelerated motion 6
1.8 Free fall under gravity 9
1.9 Angular velocity and speed 9
1.10 Angular acceleration 10
1.11 Relationship between linear speed and angular speed 10
1.12 Relationship between angular speed and frequency of rotation 11
1.13 Relationship between linear acceleration and angular acceleration 11
1.14 Uniform circular motion – acceleration 12
Exercises 14
Problems 15

Part 2 Kinetics

Chapter 2 Mass, force and acceleration
Learning objectives 21
2.1 Newton's first law of motion 21
2.2 Mass 22
2.3 Newton's second law of motion 22

2.4 Mass and weight 24
2.5 Newton's third law of motion 25
2.6 Motion of connected masses 27
2.7 Friction 29
2.8 Coefficients of friction 29
2.9 Circular motion – banked roads or tracks 31
Exercises 34
Problems 35

Chapter 3 Work, energy and power
Learning objectives 38
3.1 Work done by a constant force 38
3.2 Work done by a variable force 39
3.3 Trapezoidal rule 40
3.4 Mid-ordinate rule 40
3.5 Simpson's rule 41
3.6 Energy 42
3.7 Potential energy 42
3.8 Formula for (gravitational) potential energy 43
3.9 Kinetic energy 43
3.10 Formula for kinetic energy 44
3.11 Kinetic energy and work done 44
3.12 Conservation of energy 45
3.13 Power 46
3.14 Moment, couple and torque 48
3.15 Work done by a constant torque 49
3.16 Power transmitted by a constant torque 50
3.17 Work done by a variable torque 51
3.18 Power transmitted by belts 53
Exercises 55
Problems 58

Chapter 4 Impulse and impact
Learning objectives 62
4.1 Impulse of a force 62
4.2 Impulsive force 63
4.3 Impact of water on a fixed surface 65
4.4 Impact of water on a moving surface 66
4.5 Conservation of linear momentum 67
4.6 Impact of inelastic bodies 67
4.7 Impact of elastic bodies 69
Exercises 72
Problems 73

Part 3 Mechanics of fluids

Chapter 5 Fluids at rest
Learning objectives 77
5.1 Fluids 77
5.2 Molecular structure of fluids 78
5.3 Fluid properties 78
5.4 Pressure at a point in a fluid 82
5.5 Variation of pressure in a static fluid 83
5.6 Equality of pressures in all directions at a point 84
5.7 Pressure head 85
5.8 Gauge and absolute pressures 86
5.9 The mercury barometer 87
5.10 Centre of pressure and centroid 88
5.11 Hydrostatic thrust and position of centre of pressure for a rectangular plane surface 89
5.12 Second moment of area 90
5.13 Formulae for hydrostatic thrust and position of centre of pressure – plane immersed surface 91
5.14 Parallel axis theorem 92
Exercises 98
Problems 100

Chapter 6 Fluids in motion
Learning objectives 103
6.1 Steady flow and uniform flow 103
6.2 Laminar flow and turbulent flow 104
6.3 Streamlines 104
6.4 Continuity equation 105
6.5 The Bernoulli equation 108
6.6 Loss of pressure head in pipes 111
6.7 Reynolds number for pipe flow 111
6.8 The Darcy – Weisbach formula 111
6.9 Flow in open channels – the Chezy formula 115
6.10 Reynolds number for channel flow 117
Exercises 118
Problems 120

Chapter 7 Fluid measurements
Learning objectives 122
7.1 Pressure measurement – the piezometer 122
7.2 The simple U-tube manometer 123
7.3 The differential manometer 127

7.4　Inverted differential manometer　129
7.5　The inclined tube manometer　131
7.6　Flow measurement in pipes – the Venturi meter　132
7.7　Flow measurement in channels – notches and weirs　134
7.8　The rectangular sharp-crested weir　135
7.9　The triangular sharp-crested weir　137
7.10　The broad-crested weir　138
Exercises　139
Problems　141

Part 4 Structural mechanics

Chapter 8 Pin-jointed frameworks
Learning objectives　147
8.1　Addition of forces – the triangle rule – resultant　147
8.2　The parallelogram rule　148
8.3　Resolution of a force　148
8.4　Resultant of more than two coplanar forces – the polygon rule　149
8.5　Equilibrium and equilibrant　151
8.6　Conditions for the equilibrium of coplanar concurrent forces　152
8.7　Moment of a force　152
8.8　General conditions for the equilibrium of coplanar forces　152
8.9　Triangle and polygon of forces　153
8.10　Framed structures – types of joint　153
8.11　Types of support　154
8.12　Static determinancy　154
8.13　Pin-jointed plane frameworks　155
8.14　Conditions for statistically determinate frames　155
8.15　Analysis of statically determinate pin-jointed frameworks　157
8.16　Force diagram method　157
8.17　Semi-graphical method　164
8.18　Cantilever frameworks　169
8.19　Inclined loading　173
8.20　Combined vertical and inclined loading　179
Exercises　182
Problems　183

Chapter 9 Properties of sections
Learning objectives　188
9.1　Area and centroid　188
9.2　Neutral plane and neutral axis　191
9.3　Second moment of area　194

9.4   Section modulus   195
9.5   Radius of gyration   196
9.6   Calculation of values for second moment of area, radius of gyration and section modulus   197
9.7   Parallel axis theorem   198
Exercises   206
Problems   207

Chapter 10 Simply-supported beams
Learning objectives   211
10.1   Bending moment   211
10.2   Bending moment sign convention   212
10.3   Shearing force   212
10.4   Shearing force sign convention   213
10.5   Shearing force and bending moment diagrams   214
10.6   The four standard cases   214
10.7   Drawing a parabola   217
10.8   SF and BM diagrams – points to remember   218
10.9   Position of maximum bending moment   222
10.10   Simply supported beam with overhanging ends   225
10.11   Moment of resistance and simple beam design   230
Exercises   237
Problems   239

Chapter 11 Soil retaining walls
Learning objectives   243
11.1   Pressure due to retained granular materials   243
11.2   Trench excavation supports   246
11.3   Excavations more than 6 m deep   247
Exercises   251
Problems   251

Part 5 Answers

Objective type exercises – Chapters 1 – 11   255
Problems – Chapter 8   256
Problems – Chapter 10   265

Index   274

Preface

This text has been written primarily to cover the topics included in BTEC unit Applied Mechanics III U81/828. Very little previous knowledge of the subject is assumed and hence some material not specifically mentioned in the objectives of U81/828 is included where relevant. Each chapter includes a number of worked examples. At the end of each chapter there are multiple choice exercises and numerical problems. Answers to these are given at the back of the book.

The author acknowledges the helpful advice given by the Civil Engineering staff at Birmingham Polytechnic and the assistance of Christine Haddow who typed the manuscript.

W.M. July 1984

Part 1

Kinematics

Chapter 1

Velocity and acceleration

Learning objectives

After reading this chapter and working through the exercises, you should be able to:
- solve problems using the equations

$s = \frac{1}{2}(u+v)t$
$v = u + at$
$v^2 = u^2 + 2as$
$s = ut + \frac{1}{2}at^2$

- explain what s, u, v, a and t represent;
- define angular velocity and angular acceleration and solve problems;
- relate the above quantities to linear velocity and linear acceleration and solve problems

1.1 Scalars and vectors

A scalar quantity is one which can be completely specified by a number expressing its magnitude in an appropriate unit. Examples of scalar quantities are mass, length, area.

A vector quantity requires a number and a direction to specify it completely; that is, a vector has magnitude and direction. Examples of vectors are velocity, acceleration, force.

1.2 Linear displacement and distance

The linear displacement is the length moved in a given direction – it is a vector quantity. The magnitude of the displacement is usually known as the distance gone or simply as distance.

1.3 Linear velocity and speed

The linear velocity of a body is the rate of change of displacement with time. It is a vector quantity. A suitable unit could be the m/s.

The magnitude of the velocity is known as the speed. It is the rate of change of distance with time and is a scalar quantity.

If a body has uniform velocity, then it must move in a fixed direction with constant speed. The average speed of a body is the total distance moved divided by the total time taken to cover this distance.

1.4 Linear acceleration

The linear acceleration of a body is the rate of change of linear velocity with time. It is a vector quantity. A suitable unit could be the m/s^2. If the acceleration is uniform, the speed must be increasing by equal amounts in equal time intervals.

Worked example 1.1
A car, travelling along a straight road at a steady 13 m/s, accelerates uniformly for 15 s until it is moving at 25 m/s. Find its acceleration.

Solution

$$\text{Acceleration} = \frac{\text{change in velocity}}{\text{time taken}}$$
$$= \frac{25 - 13}{15}$$
$$= 0.8 \text{ m/s}^2$$

1.5 Distance–time curve

When a graph is plotted for a body of distance s against time t, the curve might be as shown in Fig. 1.1.

This curve can be used to deduce certain facts about the motion. Since speed is rate of change of distance with time, the slope or gradient of the s/t curve gives the speed. Over the linear section OA of

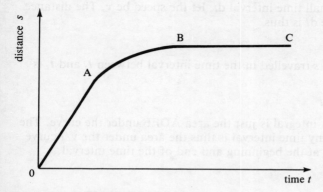

Fig. 1.1 Distance against time curve

the curve, the speed must be uniform. Between A and B, the gradient is becoming less and hence the body is slowing down. At B, the body is at rest and it remains at rest between B and C.

1.6 Speed–time curve

A graph of speed v of a body plotted against time t might be as shown by the curve in Fig. 1.2.

The acceleration of a body has magnitude equal to the rate of change of speed with time. The slope or gradient of the v/t curve is thus the acceleration. In Fig. 1.2 it follows that between A and B the acceleration is increasing, between B and C it is constant and between C and D it is decreasing.

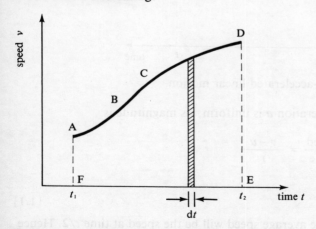

Fig. 1.2 Speed against time curve

During the small time interval dt, let the speed be v. The distance ds covered in time dt is thus

$$ds = v\, dt$$

The total distance s travelled in the time interval between t_1 and t_2 is thus

$$s = \int ds = \int_{t_1}^{t_2} v\, dt$$

However, this integral is just the area ADEF under the curve. The distance gone in any time interval is thus the area under the v/t curve between the times at the beginning and end of the time interval.

1.7 Equations for linear, uniformly accelerated motion

Suppose a body moving in a straight line has initial speed u and that it undergoes uniform acceleration a for time t. Further, let the final speed be v and the distance travelled in the time t be s. The speed–time curve will then be as in Fig. 1.3.

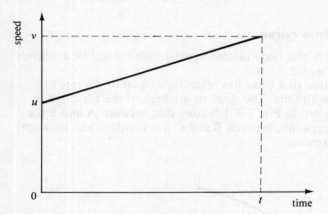

Fig. 1.3 Uniformly-accelerated linear motion

Since the acceleration a is uniform, its magnitude is

$$a = \frac{\text{change in speed}}{\text{change in time}} = \frac{v-u}{t}$$

or $at = v - u$

or $v = u + at$ [1.1]

In this case, the average speed will be the speed at time $t/2$. Hence
average speed $= \tfrac{1}{2}(u+v)$

Further, since distance gone = average speed $\times t$
then $s = \frac{1}{2}(u+v)t$ [1.2]

Substituting for v from [1.1] into [1.2] gives
$s = \frac{1}{2}(u+u+at)t$
or $s = ut + \frac{1}{2}at^2$ [1.3]

Substituting for t from [1.1] into [1.2] gives
$s = \frac{1}{2}(u+v)(v-u)/a$
or $2as = v^2 - u^2$
or $v^2 = u^2 + 2as$ [1.4]

Equations [1.1], [1.2], [1.3] and [1.4] all contain four unknowns. Three of these must be known before the fourth can be found.

Worked example 1.2
A car starts from rest and accelerates in a straight line at 1.6m/s² for 10s. What is its final speed and how far has it travelled in this time? The brakes are then applied and the car is brought to rest in 20m. Find the retardation.

Solution
(a) For the first part of the motion
Initial speed $u = 0$
Acceleration $a = 1.6$m/s²
Time $t = 10$s
Final speed $v = u + at$
Thus $v = 0 + 1.6 \times 10$
or $v = 16$m/s
Distance gone $s = ut + \frac{1}{2}at^2$
Thus $s = 0 + 0.8 \times 100$
or $s = 80$ m

(b) For the second part of the motion
Initial speed $u = 16$ m/s
Final speed $v = 0$
Distance $s = 20$ m
$v^2 = u^2 + 2as$
Thus $0 = 16^2 + 2 \times a \times 20$
or $a = -6.4$ m/s²

Now retardation is a negative acceleration. So the retardation is 6.4 m/s².

Worked example 1.3

A hoist starts at ground level and accelerates at 1.2 m/s² for 5 s. It then moves with uniform speed for 10 s and is finally brought to rest at the top of a building with retardation 2.0 m/s². Draw a speed–time sketch for the motion and find the height of the building.

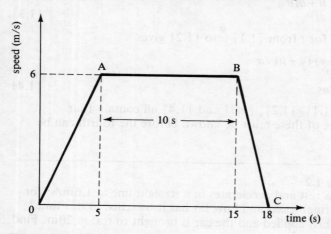

Fig. 1.4 Worked example 1.3

Solution

(a) For the first part of the motion
Initial speed $u = 0$
Acceleration $a = 1.2 \text{ m/s}^2$
Time $t = 5 \text{ s}$
Final speed $v = u + at$
   $= 0 + 1.2 \times 5$
   $= 6.0 \text{ m/s}$

(b) For the final part of the motion
Initial speed $u = 6.0 \text{ m/s}$
Final speed $v = 0$
Acceleration $a = -2.0 \text{ m/s}$ – note the negative sign
Using $v = u + at$
   $0 = 6.0 - 2t$
Thus $t = 3 \text{ s}$

The speed–time sketch is as in Fig. 1.4.
Total distance gone is the area under the speed–time sketch.
Total distance = area of trapezium OABC
   $= \tfrac{1}{2}(10 + 18)6$
   $= 84 \text{ m}$
and thus the height of the building is 84 m.

1.8 Free fall under gravity

When a body is allowed to fall freely, its acceleration is called the acceleration due to gravity g. Provided that air resistance is negligible, all bodies, heavy or light, fall with the same acceleration. The value of g varies slightly with position on the earth's surface and with altitude. However, a value of g of $9.81\,\text{m/s}^2$ is normally used in calculations and this is the value which will be used in the rest of this book.

Worked example 1.4
A workman drops a hammer from the top of a scaffolding. If the speed of sound in air is $340\,\text{m/s}$, how long does the workman have before shouting to another workman $60\,\text{m}$ vertically below him if his warning is to arrive before the hammer. Neglect air resistance.

Solution
For the hammer
Initial speed $= 0$
Acceleration $= 9.81\,\text{m/s}^2$
Distance $= 60\,\text{m}$
$$s = ut + \tfrac{1}{2}at^2$$
$$60 = 0 + \tfrac{1}{2}(9.81)t^2$$
$$t = 3.50\,\text{s}$$

The hammer takes $3.50\,\text{s}$ to fall $60\,\text{m}$. The sound take $60/340 = 0.18\,\text{s}$ to travel the same distance. So the workman has $(3.50 - 0.18) = 3.32\,\text{s}$ before shouting if the sound is to arrive before the hammer.

1.9 Angular velocity and speed

Consider a point P moving along a line QP as shown in Fig. 1.5.

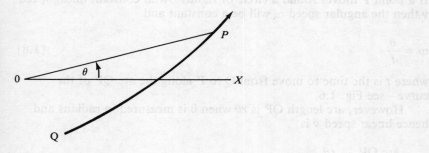

Fig. 1.5 Angular motion

Let OX be a fixed line and θ be the angle made with OP at some time t. The angular velocity ω of P about O is the rate of change of θ with time in the sense of increasing θ.

Hence $\omega = \dfrac{d\theta}{dt}$ [1.5]

When the direction of increase of θ is not included, $d\theta/dt$ is called the angular speed.

Angular speed may conveniently be measured in rad/s.

1.10 Angular acceleration

If the angular velocity of the point P in Fig. 1.5 is changing with time, then the angular acceleration α of P is the rate of change of its angular velocity, that is,

$$\alpha = \dfrac{d\omega}{dt} \qquad [1.6]$$

in the sense of increasing θ.

Angular acceleration may be measured in rad/s².

If the angular acceleration is uniform, then its magnitude is

$$\dfrac{\omega_2 - \omega_1}{t} \qquad [1.7]$$

if the angular speed changes from ω_1 to ω_2 in time t.

1.11 Relationship between linear speed and angular speed

If a point P moves round a circle of radius r with constant linear speed v then the angular speed ω will be a constant and

$$\omega = \dfrac{\theta}{t} \qquad [1.8]$$

where t is the time to move from Q to P along the arc QP of the curve – see Fig. 1.6.

However, arc length QP is $r\theta$ when θ is measured in radians and hence linear speed v is

$$v = \dfrac{\text{arc QP}}{t} = \dfrac{r\theta}{t} \qquad [1.9]$$

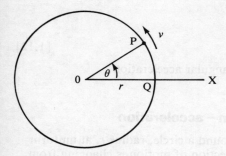

Fig. 1.6 Circular motion

Using [1.8] and [1.9] leads to

$v = r\omega$ for circular motion.   [1.10]

or linear speed = radius × angular speed

1.12 Relationship between angular speed and frequency of rotation

Let P in Fig. 1.6 rotate with constant frequency n rev/s. Since for each revolution the angle turned through is 2π rad, clearly the number of radians turned through by P every second will be $2\pi n$. Since this is the angular speed ω, then

$\omega = 2\pi n$ [1.11]

or angular speed = 2π × revolutions per second.

1.13 Relationship between linear acceleration and angular acceleration

By [1.6] and [1.10]

$$\alpha = \frac{d\omega}{dt}$$

and $v = r\omega$

Hence $\alpha = \dfrac{d}{dt}\left(\dfrac{v}{r}\right) = \dfrac{1}{r}\dfrac{dv}{dt}$ since r is constant

However dv/dt is linear acceleration a

Hence $\alpha = \dfrac{a}{r}$

or $\quad a = r\alpha \quad\quad$ [1.12]

or linear acceleration = radius × angular acceleration.

1.14 Uniform circular motion – acceleration

Consider again a point P moving round a circle, radius r, at uniform speed v as in Fig. 1.6. Since the direction of motion is changing from instant to instant, the velocity is changing and hence the point P has acceleration. Let AB and AC represent its velocity in magnitude and direction when it is at Q and P respectively – see Fig. 1.7. AB is parallel to the tangent at Q and AC is parallel to the tangent at P.

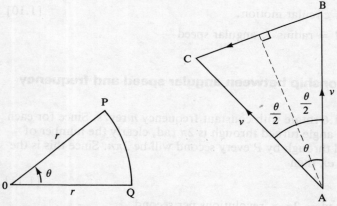

Fig. 1.7 Angular acceleration

The line BC represents the change in velocity as the point moves from Q to P.

Average acceleration of P is

$$a = \dfrac{\text{length of BC}}{\text{time to go from Q to P}}$$

Now length of arc QP = $r\theta$ and hence time to go from Q to P is $r\theta/v$.

Thus $a = \dfrac{BC \times v}{r\theta}$

However, BC = $2v \sin \theta/2$

Hence $a = \dfrac{v^2}{r} \dfrac{\sin \theta/2}{\theta/2}$

To find the acceleration of P at Q, let the angle $\theta/2$ tend to zero. Then $\sin \theta/2 \to \theta/2$ and the acceleration a becomes

$$a = \dfrac{v^2}{r} \qquad [1.13]$$

When $\theta/2 \to 0$, the direction of BC approaches the direction QO, that is, towards the centre of the circle.

Hence the acceleration of a point moving round a circle with radius r at constant speed v is v^2/r towards the centre of the circle.

Worked example 1.5

A flywheel rotating at 1200 rev/min slows down at a constant rate to 900 rev/min in 30 s. Find:

(i) the initial angular speed,
(ii) the final angular speed,
(iii) the angular acceleration,
(iv) the initial speed of a point on the rim of the flywheel if its diameter is 1.1 m.

Solution

(i) Angular speed $\omega = 2\pi n$
Initial $n = 1200$ rev/min $= 1200/60 = 20$ rev/s
Hence initial angular speed $= 2\pi \times 20 = 125.7$ rad/s
(ii) Final $n = 900$ rev/min $= 900/60 = 15$ rev/s
Hence final angular speed $= 2\pi \times 15 = 94.2$ rad/s
(iii) Change in angular speed $= 94.2 - 125.7$
or $\omega_2 - \omega_1 = -31.5$ rad/s
Angular acceleration $\alpha = (\omega_2 - \omega_1)/\text{time}$

$$\alpha = -\dfrac{31.5}{30}$$
$$= -1.05 \text{ rad/s}^2$$

(iv) Linear speed $v = r\omega$
Hence initial speed of a point on the rim is
$v = \tfrac{1}{2} \times 1.1 \times 125.7$
$= 69.1$ m/s

Worked example 1.6
The spin drier in a washing machine is a cylinder with diameter 500mm. It spins at 900rev/min. Find the speed and acceleration of a point on the side of the drum.

Solution
Angular speed $\omega = 2\pi n$
$n = 900/60 = 15\,\text{rev/s}$
Thus $\omega = 2\pi \times 15 = 94.2\,\text{rad/s}$
Speed $v = r\omega$
$\qquad = \frac{1}{2} \times 0.5 \times 94.2$
$\qquad = 23.6\,\text{m/s}$
Acceleration $= v^2/r$
$\qquad\qquad = 23.6^2 / 0.25$
$\qquad\qquad = 2230\,\text{m/s}^2$

Note that this is 227 times the acceleration due to gravity.

Exercises

Objective type
Choose the ONE response which is the most appropriate.

1.1 Which of the following is a vector quantity?
    A   density
    B   speed
    C   area
    D   acceleration

1.2 Velocity is rate of change with time of
    A   displacement
    B   acceleration
    C   speed
    D   distance

1.3 A unit used for acceleration is the
    A   m/s
    B   m/s^2
    C   m^2/s
    D   ms^2

1.4 The speed of a car moving in a straight line increases uniformly from 8m/s to 14m/s in 3s. The acceleration, in m/s^2, is
    A   2
    B   11/3
    C   6
    D   22/3

1.5 A stone is projected vertically upwards with an initial speed of 20m/s. After one second its acceleration, in m/s², is
  A zero
  B 4.90 upwards
  C 9.81 upwards
  D 9.81 downwards

1.6 The sketch shows how the speed of a car varies with time. The acceleration of the car, in m/s², is

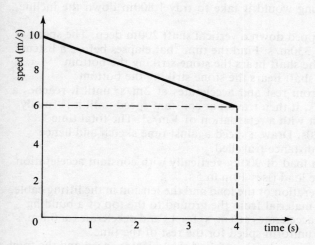

Fig. 1.8 Exercise 1.6

  A −1
  B 1
  C −2.5
  D 2.5

1.7 When a body moves round a circle with radius r at uniform speed v, the angular speed ω is
  A vr
  B v/r
  C v^2/r
  D $2\pi v$

1.8 What is the angular speed, in rad/s, of a body moving in a circle at 300 rev/min?
  A 600π
  B 300π
  C 100π
  D 10π

Problems

1.9 A man swims at a steady 1 m/s across a lake and then swims back at a steady 0.5 m/s. What was his average speed?

1.10 A car travelling at 50 km/h stops in 11 m. Find its retardation, assuming it is uniform.

1.11 A wagon starts to run, from rest, down an incline. In 30 s it travels 100 m. How long would it take to travel 800 m down the incline, from rest?

1.12 A stone is dropped down a vertical shaft 200 m deep. The speed of sound in air is 330 m/s. Find the time that elapses before a listener at the top of the shaft hears the stone striking the bottom.
the top of the shaft hears the stone striking the bottom.

1.13 A lift starts from rest and accelerates at 2 m/s^2 until it reaches a speed of 3 m/s. It then travels at constant speed until it is finally brought to rest with a retardation of 1 m/s^2. The total time of motion is 20 s. Draw a speed against time sketch and hence find the total distance travelled.

1.14 A crane lifts a load of 900 kg vertically with constant acceleration from rest. The load rises 10 m in 5 s.
Find the acceleration of the load and the tension in the lifting cable.

1.15 A hoist raises material from the ground to the top of a building 100 m high. It accelerates at 1 m/s^2 for 1 s and decelerates at 0.8 m/s^2 and travels at uniform speed for the rest of the time.
Draw a speed–time sketch and find the uniform speed and the total time of travel from the ground to the top of the building.

1.16 A train accelerates and decelerates at 0.5 m/s^2. What is the minimum time it can take to travel between two stations 5 km apart?

1.17 A man gets out of a train and walks along the platform at 2 m/s. The train begins to move 30 s later and accelerates at 0.5 m/s^2. How much further time will elapse before the door from which the man left the train overtakes him?
train overtakes him?

1.18 A pulley with diameter 1 m rotates 120 times per minute. It has a belt passing round it. Find
(a) the angular speed of the pulley;
(b) the linear speed of the belt.

1.19 Find the linear speed of a point on the earth's equator assuming that the radius of the earth is 6380 km.

1.20 A flywheel rotates at 300 rev/min. It slows uniformly to 180 rev/min in 30 s. What is its angular acceleration?

1.21 A grinding wheel is accelerated uniformly from rest to 3000 rev/min in 3.0 s. Find its angular acceleration. If the wheel diameter is 200 mm, find the final linear speed of a point on its rim.

1.22 A train on a level straight track accelerates uniformly from rest to 60 km/h in a distance of 300 m. Find the acceleration. If the driving wheels are 1.5 m in diameter, find their angular acceleration and the final angular speed of a point on the rim of a wheel.

Part 2

Kinetics

Chapter 2

Mass, force and acceleration

Learning objectives

After reading this chapter and working through the exercises, you should be able to:
- state Newton's first law of motion;
- state Newton's second law of motion in terms of $P = ma$;
- state Newton's third law of motion;
- solve problems involving Newton's laws and including connected masses.

In the last chapter motion was dealt with without any consideration of the forces involved. In this chapter, the relationship between mass, force and acceleration will be explored.

2.1 Newton's first law of motion

This law is really only a definition of force. It says that when a body is acted on by a resultant external force, it will accelerate. If the resultant external force is zero, then the body will either remain at rest or else it will continue to move with constant velocity, that is, with constant speed in a straight line.

To illustrate this law, consider the example of a parachutist jumping from an aeroplane flying horizontally. His initial vertical speed will be zero. However, he will be acted on by his weight acting vertically downwards and since the external resultant force is not zero, he will accelerate. He will not, however, increase in speed indefinitely because as his speed increases, so does the frictional drag. Since this opposes his motion, eventually weight and drag will be equal and opposite. This means that the resultant external force is now zero and a constant or terminal speed will be attained. For the human body falling freely this constant speed is of the order of 45–55 m/s or 100–120 mph.

If the parachutist should now release his parachute, frictional drag will suddenly be greatly increased and will exceed the weight. There is again a resultant external force which is now upwards and hence there will be retardation. This will continue until drag and weight are again equal and opposite and when this happens the parachutist will have constant speed once more. The value of this will be only a few m/s.

The first law can be summarized as 'A body not acted on by an external resultant force moves with constant velocity or is at rest'.

2.2 Mass

The mass of a body is sometimes said to be the quantity of matter in the body. This is not a satisfactory statement because of the vagueness of the word quantity.

The mass of a body is a number assigned to it to distinguish it from another which may appear to be identical. It determines the behaviour of the body when it is acted on by a force which causes it to change its motion. The mass can be considered to be a measure of the resistance to change of motion. For example, if two rollers, one large and one small, but similar in all other respects, are pushed along a horizontal lawn, it will soon be obvious that the smaller one is easier to start and stop moving. This is because the smaller one has a smaller mass. The resistance to change of motion is known as **inertia**. A body with large mass is said to have large inertia. The first law of motion is sometimes called the 'law of inertia'.

The mass of a body is a constant for all cases where the velocity is very small compared with the velocity of light, namely 3×10^8 m/s. However, for velocities which are a substantial fraction of the velocity of light, the increase of mass with velocity must be taken into account. This is one aspect of the branch of mechanics called **Relativistic Mechanics**. The SI unit of mass is the kg.

The relationship between mass and acceleration is summed up in Newton's second law of motion.

2.3 Newton's second law of motion

Before stating this law, linear momentum, usually simply called momentum, must be defined. This is a very important quantity in mechanics. It is denoted by p and is defined as

momentum = mass × velocity
or $\quad p = mv$ [2.1]

Momentum is a vector quantity and may be measured in kg m/s or in Ns. The reader should verify that these two units are, in fact, equivalent.

The second law states that the force causing acceleration is proportional to rate of change of momentum with time and acts in the direction of the change.

Now if a force P changes the velocity of a body with constant mass m uniformly from u to v in time t, then the rate at which momentum changes is $(mv - mu)/t$.

Thus by Newton's second law

Force $P \propto \dfrac{m(v-u)}{t}$

or $\quad P = k\,ma$

since $(v-u)/t$ is acceleration a. In this equation k is a constant of proportionality. By definition, 1 N causes an acceleration of $1\,\text{m/s}^2$ of a mass of 1 kg. This gives $k = 1$ and hence the second law of motion may be summarized as

'Accelerating force = mass × acceleration'
or $P = ma$ [2.2]

with P and a both measured in the same direction.

Newton's first law is a special case of his second law with the accelerating force P zero.

Worked example 2.1
An engine pulls a wagon of mass 10 000 kg along a straight track at a steady speed. The pull in the coupling between the wagon and engine is 1000 N. What is the force opposing the motion of the wagon? If the pull in the coupling was increased to 1200 N and the resistance to the motion of the wagon remained constant, what would be the acceleration of the wagon?

Solution
When the speed is steady, by Newton's first law, the resultant external force on the body is zero. The pull exerted on the wagon by the coupling must be exactly balanced by the resistance to motion and hence the force opposing motion is 1000 N.

When the pull in the coupling is 1200 N the accelerating force P is $1200 - 1000 = 200\,\text{N}$.

Since $P = ma$
$200 = 10\,000\,a$
Hence acceleration $a = 0.02\,\text{m/s}^2$

Worked example 2.2

Find the acceleration of a 20 kg crate along a horizontal floor when it is pushed with a resultant force of 10 N parallel to the floor. How far will the crate move in 5 s from rest?

Solution

Accelerating force $P = 10\,\text{N}$

$P = ma$
$10 = 20\,a$
$a = 0.5\,\text{m/s}^2$

Distance $s = ut + \tfrac{1}{2}at^2$
$s = 0 + \tfrac{1}{2}(0.5)(5)^2$
$s = 6.25\,\text{m}$

2.4 Mass and weight

The weight of a body may be defined as the force with which it is attracted towards the earth. Now when a body falls freely (strictly speaking in a vacuum) its acceleration is constant. This constant acceleration is denoted by g and is called the acceleration due to gravity. Since the accelerating force is mg the second law of motion gives

Weight $W = mg$ [2.3]

If m is in kg and g is in m/s^2, W will be in N.

Since g varies slightly over the surface of the earth the weight of a body at rest is not a constant as is the mass, but may vary slightly from place to place.

Worked example 2.3

A 1 kg stone falls freely, from rest, from a bridge. What is the force causing it to accelerate? What is its speed 4 s later and how far will it have fallen in this time?

Solution

The force causing the stone to accelerate is its weight W where

$W = mg = 1 \times 9.81 = 9.81\,\text{N}$

The acceleration of the stone will be $9.81\,\text{m/s}^2$ and its initial speed is zero.

Then using

$$v = u + at$$

after 4s its speed will be

$$v = 0 + 9.81 \times 4$$
$$= 39.2 \, \text{m/s}$$

The distance fallen can be found using

$$s = ut + \tfrac{1}{2} at^2$$
$$= 0 + \tfrac{1}{2}(9.81)(4)^2$$
$$= 78.5 \, \text{m}$$

2.5 Newton's third law of motion

This law states that if a body A exerts a force F on body B, then body B exerts an equal but opposite force on body A. For example, when a block with mass M rests on a horizontal table, the mass exerts a downward force on the table equal to its weight Mg. By Newton's third law, the table exerts a force of Mg vertically upwards on the block. The third law applies to bodies at rest or in motion.

Consider the example of a truck towing a car by means of a rope. If, at any instant, the forward force exerted by the rope on the car is T, then at that instant, by the third law, the rope exerts a backward force T on the truck. The force T is known as the tension in the rope.

Newton's third law is sometimes summarized as 'Every action has an equal and opposite reaction'.

Worked example 2.4
A lift with its load has a total mass of 2000 kg. It is supported by a steel cable. Find the tension in the cable when the lift

(i) is at rest,
(ii) accelerates upwards with uniform acceleration $1 \, \text{m/s}^2$,
(iii) moves upwards with steady speed 1 m/s,
(iv) moves downwards with steady speed 1 m/s,
(v) accelerates downwards with uniform acceleration $1 \, \text{m/s}^2$.

Solution
(i) When the lift is at rest, by Newton's first law the resultant force on it is zero. Since the two forces acting on the lift are its weight $W = mg$ acting vertically downwards and tension T in the cable (Fig. 2.1) acting vertically upwards then

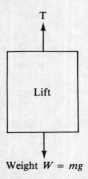

Fig. 2.1 Worked example 2.4

$$T = mg$$
$$= 2000 \times 9.81$$
$$= 19\,600\,\text{N}$$
$$= 19.6\,\text{kN}$$

(ii) If the lift accelerates upwards T must exceed mg and the resultant accelerating force $P = T - mg$
But $P = ma$ by Newton's second law.
Thus $T - 2000g = 2000 \times 1$
or $\quad\quad\quad T = 2000 + 19\,600$
$\quad\quad\quad\quad = 21\,600\,\text{N}$
$\quad\quad\quad\quad = 21.6\,\text{kN}$

(iii) As in (i), by Newton's first law, the resultant force on the lift must be zero and hence

$$T = mg$$
$$= 19.6\,\text{kN}$$

(iv) As in (iii) the tension in the cable will still equal mg since the change in direction of motion cannot alter the fact that there is no acceleration and hence

$$T = 19.6\,\text{kN}$$

(v) If the lift accelerates downwards, then mg must exceed T and the resultant accelerating force $P = mg - T$
But $P = ma$ by Newton's second law.
Thus $2000 \times 1 = 2000\,g - T$
or $T = 19\,600 - 2000$
$\quad\quad = 17\,600\,\text{N}$
$\quad\quad = 17.6\,\text{kN}$

2.6 Motion of connected masses

In this section, the motion of two masses connected by a light, inextensible cable or rope passing over a light pulley with frictionless bearings will be considered. Under these conditions, the tension in the rope is the same throughout its length. Also, the magnitudes of the velocities of the two masses and of the accelerations of the two masses must be equal.

Consider the motion of two masses m_1 and m_2 connected by a light, inextensible rope passing over a light pulley as shown in Fig. 2.2 and let m_1 be greater than m_2.

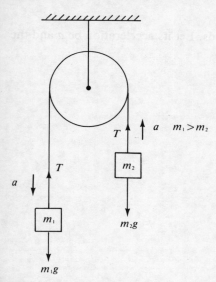

Fig. 2.2 Connected masses

Suppose m_1 accelerates downwards with acceleration a. m_2 will thus accelerate upwards with acceleration a. Let the tension throughout the rope be T. T and a can be found by applying Newton's second law to m_1 and to m_2.

The forces acting on m_1 are its weight m_1g acting vertically downwards and the tension T in the rope acting vertically upwards. If m_1 moves downwards, m_1g must exceed T and the accelerating force P on m_1 is $(m_1g - T)$

Since $P = m_1a$
$$m_1g - T = m_1a \qquad [2.4]$$

The forces acting on m_2 are its weight m_2g downwards and tension T upwards. If m_2 moves upwards, T must exceed m_2g and the accelerating force P on m_2 is $(T - m_2g)$.

Since $P = ma$

$$T - m_2 g = m_2 a \qquad [2.5]$$

Equations [2.4] and [2.5] can now be solved simultaneously for T and a. A worked example will illustrate the method.

Worked example 2.5

Masses of 1.2 kg and 1.0 kg hang at the ends of a light rope passing over a light, frictionless pulley. Find the acceleration of the masses and the tension in the string.

Solution:

The 1.2 kg mass will move downwards. Let its acceleration be a and the tension in the rope be T (Fig. 2.3).

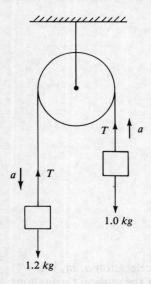

Fig. 2.3 Worked example 2.5

For the 1.2 kg mass, accelerating force $P = 1.2\,g - T$
Thus $1.2\,g - T = 1.2\,a \ldots$ [1]
For the 1.0 kg mass, accelerating force $P = T - 1.0\,g$
Thus $T - 1.0g = 1.0\,a \ldots$ [2]
Addition of [1] and [2] gives
$1.2g - 1.0g = 2.2\,a$
or $a = 0.2 \times 9.81/2.2$
    $= 0.892 \text{m/s}^2$

Substitution for a into [2] gives

$T = 1.0 \times 9.81 + 1.0 \times 0.892$
$\quad = 10.7 \text{N}$

2.7 Friction

The force which prevents or tries to prevent the slipping or sliding of two surfaces in contact is called friction. Several rules or laws have been developed as a result of experiment and experience. These laws apply only to dry surfaces. They can be stated as follows.

1. The frictional resistance between two sliding surfaces is directly proportional to the force pressing the two surfaces together.

   This law breaks down when the force or reaction between the surfaces is very small or very large. For very large forces seizing of the surfaces may occur.

2. The frictional resistance depends on the nature and roughness of the surfaces involved.

3. The frictional resistance is independent of the area of the surfaces in contact.

   This law breaks down if the area becomes so small that damage of the surface occurs leading to an increase in the frictional resistance.

4. The frictional resistance is independent of the speed of sliding.

   This law breaks down when the speed of sliding is very high or very low. When the speed of sliding is very high, the temperature of the surfaces may increase and this may lead to seizing of the surfaces. The frictional resistance is greatest when the speed of sliding is zero and when motion is just about to commence.

From the remarks following the above laws, it should be clear that the laws must be used with caution.

The laws for lubricated surfaces differ widely from those stated above for dry surfaces.

2.8 Coefficients of friction

Consider a block with mass m resting on a horizontal surface as in Fig. 2.4.

Let the force or thrust between the surfaces be N – known often as the normal reaction. N will equal the weight mg by Newton's third law. Also, let the external force applied to the block be F_A. As F_A increases from zero, the frictional resistance to motion, F, will also increase from zero. Eventually F reaches its maximum value F_S and the block will be on the point of moving. In this situation $F_A = F_S$ and by law 1,

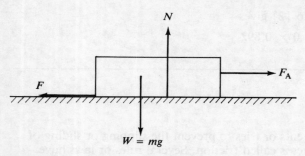

Fig. 2.4 Block on a rough surface

$$F_S \propto N$$
$$\text{or } F_S = \mu_S N \qquad [2.6]$$

where the constant μ_S is called the coefficient of static friction. When F_A exceeds F_S, the block will start to slide. It is an experimental fact that once the block starts to move, the applied force required to keep it moving steadily is less than F_S, that is, the frictional resistance F when sliding is taking place is less than F_S. This means that the coefficient of sliding friction μ defined by $F = \mu N$ is less than μ_S. [2.7]

Values of μ and μ_S depend on the surfaces in contact. Some typical values are given below:

steel on steel $\mu_S = 0.58$, masonry on rock $\mu_S = 0.6-0.75$
wood on brick $\mu_S = 0.6$, masonry on clay $\mu_S = 0.30$
natural rubber, vulcanized sliding on clean bitumen at 100m/min
$\mu = 1.07$

Worked example 2.6
A crate with mass 50kg will just slide with uniform speed down a rough ramp at 30° to the horizontal.
Find the coefficient of friction.

Solution
The forces acting on the crate are its weight mg, the normal reaction N of the ramp and the frictional resistance F. These are shown in Fig. 2.5.

Since there is no acceleration down the ramp the resultant force parallel to the ramp must be zero. Resolution of the forces parallel to the plane thus gives

$$F - \text{component of weight down ramp} = 0$$
$$\text{or } F = mg \cos 60° = 50 \times 9.81 \times 0.5$$
$$= 245\text{N}$$

Also, since there is no acceleration at right angles to the ramp, the

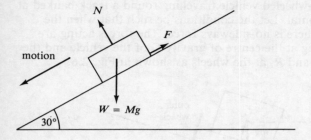

Fig. 2.5 Worked example 2.6

resultant force at right angles to the ramp must also be zero. Resolution of the forces at right angles to the ramp gives

$N = mg \cos 30° = 50 \times 9.81 \times \cos 30°$
$ = 425 \text{ N}$

But coefficient of friction

$$\mu = \frac{F}{N}$$

Thus $\mu = 245/425 = 0.576$

2.9 Circular motion – banked roads or tracks

In Chapter 1, the acceleration of a body moving round a circle, radius r, with uniform speed v was found to be v^2/r towards the centre of the circle. The force P producing this acceleration in a mass m must be

$$P = ma = \frac{mv^2}{r} \qquad [2.8]$$

by Newton's second law.

In this case the force P is called the centripetal force. The force may be provided in various ways. For example, in the case of a car travelling round a curved level road, it is the friction between the tyres and the road which provides the force.

When a curved track is banked downwards towards the centre of motion, then the frictional force in the case of a car and the sideways thrust in the case of a vehicle on rails are less than they would be if the curved track was level and the speed of motion was the same. The reduced dependence on friction or side rail thrust to provide the centripetal force gives greater safety and comfort.

Consider a four-wheeled vehicle travelling round a track banked at angle θ to the horizontal. Let the conditions be such that when the vehicle has speed v there is no sideways force. The forces acting are then the weight acting at the centre of gravity G of the vehicle and the normal reactions R_1 and R_2 at the wheels as shown in Fig. 2.6.

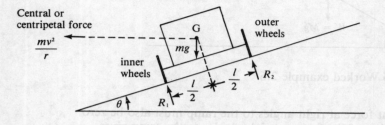

Fig. 2.6 Vehicle on a banked track

Taking moments about G gives

$$\frac{R_1 l}{2} = \frac{R_2 l}{2}$$

or $R_1 = R_2 = R$, say.

This means that the reactions at the two wheels are equal.

Vertically, there is no motion and hence resolving the forces vertically gives

$$2R \cos \theta = mg \qquad [2.9]$$

Horizontally, by Newton's second law

$$P = ma = mv^2/r$$

and hence resolving the forces horizontally gives

$$\frac{mv^2}{r} = 2R \sin \theta \qquad [2.10]$$

Using [2.9] and [2.10]

$$\tan \theta = \frac{\sin \theta}{\cos \theta} = \frac{mv^2}{mgr}$$

or $\tan \theta = \dfrac{v^2}{gr}$ \qquad [2.11]

This is the condition for no side thrust.

In the case of a rail vehicle, where the speed v is greater than that given by [2.11], there will be an outward thrust on the outer rail and when v is less than that given by [2.11], there will be an inward thrust on the inner rail. The height of the outer rail above the inner rail is known as the superelevation.

Worked example 2.7
A train travels round a banked circular track with radius 300 m at a steady speed of 84 km/h. Find the superelevation for no lateral force if the track width is 1.435 m.

Solution
The situation is illustrated in Fig. 2.6.

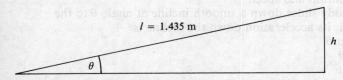

Fig. 2.7 Worked example 2.7

Vertically, $2R \cos \theta = mg$

Horizontally, $2R \sin \theta = \dfrac{mv^2}{r}$

Thus $\tan \theta = \dfrac{v^2}{gr}$

Now $v = \dfrac{84 \times 10^3}{3600}$

$= 23.3$ m/s

Thus $\tan \theta = \dfrac{23.3^2}{9.81 \times 300} = 0.184$

and $\theta = 10.4°$

The superelevation $h = l \sin \theta$ where l is the track width—see Fig. 2.7. Thus $h = 1.435 \sin 10.4° = 0.259$ m or 259 mm.

Exercises

Objective type
Choose the ONE response which is the most appropriate.

2.1 Which of the following units can be used to measure force?
    A kg
    B m/s
    C m/s^2
    D kN

2.2 Which pair of quantities are both vectors?
    A force and speed
    B mass and acceleration
    C velocity and weight
    D momentum and speed

2.3 When a body slides down a smooth incline at angle θ to the horizontal, its acceleration down the incline is
    A $mg \sin \theta$
    B $g \sin \theta$
    C $g \cos \theta$
    D $mg \cos \theta$

2.4 The force causing a body to accelerate has magnitude equal to the rate of change with time of
    A velocity
    B displacement
    C momentum
    D speed

2.5 Angular acceleration can be measured in
    A m/s^2
    B rad/s
    C rad^2/s
    D rad/s^2

2.6 A 5 kg block, at rest on a smooth horizontal surface, is acted on by a resultant force of 2.5 N parallel to the surface. The acceleration, in m/s^2, is
    A 0.5
    B 2
    C 12.5
    D 2000

2.7 A crate of mass 100 kg, is pulled along a level floor by a rope parallel to the floor. The tension in the rope is 50 N. The frictional force between crate and floor is 10 N. The acceleration of the crate, in m/s^2 is
    A 0.1
    B 0.4
    C 0.5
    D 0.6

2.8 A block can just be pulled along a rough horizontal floor by a force of 300 N parallel to the floor. The coefficient of friction between floor and block is 0.5. The mass of the block, in kg, is
A 15.3
B 61.2
C 150
D 600

2.9 A box whose mass is 50 kg is pulled at constant speed across a level floor by a force of 200 N applied at 40° to the horizontal. The friction force, in N, is
A 129
B 153
C 200
D 490

2.10 A timber plank can be made to slide, with constant speed, across a level surface by applying a force of 20 N. If the plank is now cut into two equal pieces, which are placed on top of one another, the force required to make them slide with constant speed, in N, is now
A 10
B 20
C 40
D 80

2.11 A concrete block has mass 30 kg. Its weight, in kN, is
A 0.294
B 3.06
C 30
D 294

2.12 The centripetal force, in N, on a body of mass 8 kg moving in a circle of diameter 2 m at 300 rev/min is
A 200
B 400
C $200\pi^2$
D $800\pi^2$

2.13 The angular speed of a wheel decreases uniformly from 900 rad/min to 360 rad/min in 3 minutes. Its angular retardation, in rad/s², is
A 180
B 5
C 3
D 0.05

Problems

2.14 What resultant force will cause a 2 kg mass to accelerate at 1.2 m/s²?

2.15 A resultant force of 100 N on a body produces an acceleration of 0.1 m/s². Find the weight of the body.

2.16 What force is required to accelerate a car of mass 2000 kg from rest to 50 km/h in 10 s?

2.17 The engine of a train of total mass 200 t exerts a pull of 70 kN and the resistance to motion is 15 kN. Find the acceleration of the train.

2.18 Find the time taken for a barge of mass 10 000 kg moving at 1.5 m/s to be brought to rest if the only force on it is due to a rope in which the tension is 10 kN. Find also the distance gone in this time.

2.19 A man of mass 70 kg stands in a lift. What will be the force exerted on him by the floor of the lift when
  (i) the lift is at rest,
  (ii) the lift is ascending with acceleration 1 m/s²
  (iii) the lift is descending with acceleration 1 m/s²
  (iv) the lift is moving upwards with uniform speed 2 m/s?

2.20 A train of mass 200 tonne travels with uniform speed along a straight level track. The total resistance to motion is 50 N/tonne. What is the driving force at the wheels of the engine?

2.21 If the train in Problem 2.20 now accelerates at 0.2 m/s², what is the new value of the driving force if the resistance to motion remains the same?

2.22 An electric train has total mass 300 t and the tractive force is 100 kN. If frictional forces are 25 kN and are constant, find the acceleration
  (i) along the level,
  (ii) up an incline of 1 in 150 along the incline,
  (iii) down an incline of 1 in 100 along the incline.

2.23 A 2 tonne truck is pulled at uniform speed up an incline of 1 in 10 along the slope by a rope, parallel to the incline. The rope passes over a frictionless pulley at the top of the incline and has a 275 kg mass hanging freely at its end. Find the resistance to motion parallel to the incline.

2.24 A hoist of mass 1.2 t is pulled vertically upwards, from rest, by a cable, first with uniform acceleration 1 m/s², then at uniform speed and finally with uniform retardation 2 m/s². Find the tensions in the cable during the three parts of the motion.

2.25 The normal reaction between the driving wheels of a locomotive and the track is 1000 kN and the coefficient of friction is 0.25. What is the greatest driving force which can be exerted by the locomotive parallel to the track? Find the maximum acceleration when the locomotive pulls a number of wagons along the level if the total mass of the wagons is 200 tonne and the total resistance to motion is 50 kN.

2.26 Masses of 5 kg and 4 kg are attached to the ends of a light string passing over a light, frictionless pulley. Find the acceleration of the masses and the tension in the string.

2.27 A thin, flexible wire passes over a pulley. Masses of 2.6 kg and 2.0 kg are attached to the two ends of the wire. The system is released from rest when the two masses are on the same horizontal level. Ignoring friction and the masses of the wire and pulley, find the acceleration of the masses, the tension in the wire and the time taken for the two masses to be 0.5 m apart vertically.

2.28 A car travels at a constant speed of 60 km/h round a circular track with radius 100 m banked at an angle θ to the horizontal. Find the value of θ if there is no lateral frictional force between the tyres and the track.

2.29 Calculate the height of the outer rail above the level of the inner rail on a circular curve of 200 m radius on a railway with gauge 1.435 m so that there is no side thrust when a train travels round the curve at 80 km/h.

2.30 An astronaut undergoing training is subjected to high acceleration forces by putting him in a cabin moving in a horizontal circle at the end of a whirling arm with effective length 8 m. Find the speed of rotation of the arm which would produce an acceleration of 10 g.

Chapter 3

Work, power and energy

Learning objectives

After reading this chapter and working through the exercises, you should be able to:
- describe potential energy as energy due to position, derive potential energy as mgh and calculate examples of potential energy;
- describe kinetic energy as energy due to motion, derive kinetic energy as $mv^2/2$ and calculate examples of kinetic energy;
- state the law of conservation of energy and solve problems involving situations where mechanical energy is conserved;
- define power as the rate of energy transfer and solve associated problems;
- define couple and torque and calculate work done by:
  (a) a variable force
  (b) a variable torque;
- solve problems involving loss of energy due to friction.

3.1 Work done by a constant force

When the point at which a force acts moves, the force is said to have done work. When the force is constant, the work done is defined as work done = force × distance moved in the direction of the force. It is a scalar quantity.

If a constant force F moves a body from A to B then the distance

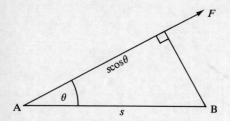

Fig. 3.1 Notation for work done

moved in the direction of F is $s \cos \theta$—see Fig. 3.1. The work done by a constant force is thus

Work done $= Fs \cos \theta$ [3.1]

If the body moves in the same direction as the force, $\theta = 0°$ and work done is Fs. Note that when $\theta = 90°$, the work done is zero. If F is in newtons and s is in metres, the work done will be measured in joules (J).

Worked example 3.1
How much work is done when a force of 5 kN moves its point of application 600 mm in the direction of the force.

Solution
Work done = force × distance
                  = $5 \times 10^3 \times 600 \times 10^{-3}$
                  = 3000 J
                  = 3 kJ

Worked example 3.2
Find the work done in raising 100 kg of water through a vertical distance of 3 m.

Solution
In this case the force is the weight of the water which is 100 g newtons.
Work done = force × distance
                  = $100 \times 9.81 \times 3$
                  = 2943 J

3.2 Work done by a variable force

Forces in practice often vary. In such cases, [3.1] cannot be used. Let the force vary as in Fig. 3.2.

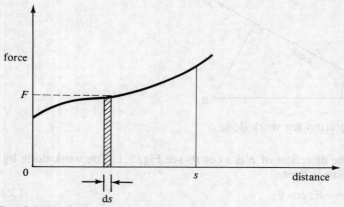

Fig. 3.2 Work done by a variable force

Consider a thin strip with width ds — shown shaded in Fig. 3.2. Over the distance ds, the force can be considered to be constant at F. The work done for the distance ds is thus F ds, which is the area of the shaded strip. The total work done for the distance s is the sum of the areas of all such strips. This is just the area under the force–distance curve. For a variable force then,

Work done = area under force–distance graph. [3.2]

If the force is constant, [3.2] can still be used. The area under the graph will be that of a rectangle with length s and breadth F, that is Fs, agreeing with [3.1]. When $\theta = 0°$.

The area under the graph can be found using the trapezoidal, mid-ordinate or Simpson's rule. These are summarized below.

3.3 Trapezoidal rule

If the area under the graph is divided into a number of vertical strips all with width w then

Total area = w × (half the sum of first and last ordinate s + sum of all other ordinates) [3.3]

3.4 Mid-ordinate rule

If the area under the graph is divided into a number of vertical strips all with width w and the ordinates at the middle of each strip are drawn then

Total area = w (sum of mid-ordinates) [3.4]

3.5 Simpson's rule

This rule is the most accurate of the three mentioned. The area under the graph must be divided into an *even* number of vertical strips each with width w. Then

Total area = $\frac{1}{3} w$ (sum of first and last ordinates + 4 times sum of even ordinates + 2 times sum of other odd ordinates) [3.5]

Worked example 3.3

The force exerted by a cable on a wagon varies as follows:

Distance moved (m)	0	20	40	60	80	100	120
Force (kN)	2.8	4.7	6.2	7.3	7.1	6.7	6.2

Find the work done on the wagon during the time it moves from 0–120 m.

Solution

The graph of force F plotted against distance s is as shown in Fig. 3.3 with six strips. To show how the rules mentioned above can be applied, the area will be found using all three.

Work done = area under force–distance graph

(a) Trapezoidal rule.

Area = w (half sum of first and last ordinates + sum of all other ordinates)

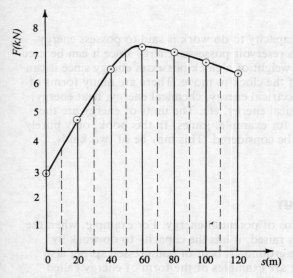

Fig. 3.3 Worked example 3.3

Thus Area = 20 [½(2.8 + 6.2) + 4.7 + 6.2 + 7.3 + 7.1 + 6.7]
= 730

and hence the work done is 730 kJ. The units will be kJ since force is in kN and distance is in m.

(b) Mid-ordinate rule.
Area = w (sum of mid-ordinates)
Thus Area = 20 (3.85 + 5.5 + 6.95 + 7.25 + 6.9 + 6.45)
= 738

and hence the work done is 738 kJ.

(c) Simpson's rule.
This can be used since there is an even number of strips.
Area = ⅓ w (sum of first and last ordinates + 4 times sum of even ordinates + 2 times sum of other odd ordinates)
Thus Area = ⅓(20)[2.8 + 6.2 + 4(4.7 + 7.3 + 6.7) + 2(6.2 + 7.1)]
= 736

and hence the work done is 736 kJ.

Using a larger number of strips would increase the accuracy of the result obtained but in practice the extra work involved would probably not be justified.

3.6 Energy

A body which has the capacity to do work is said to possess energy. For example, water in a reservoir possesses energy since it can be used to drive a turbine. The weight of a clock possesses energy since it can cause the mechanism of the clock to move. There are many forms of energy, for example, electrical energy, chemical energy, heat energy, nuclear energy, mechanical energy, etc. The units of energy are the same as those of work, for example, joules. In this book, only purely mechanical energy will be considered. This may be of two kinds, potential or kinetic.

3.7 Potential energy

There are different forms of potential energy. For example, when the driver of a pile-driver is raised, it has the capacity to do work by driving the pile into the ground. A bent or compressed spring also possesses energy. These are examples of the form of energy called potential energy. Only gravitational potential energy will be considered here. It may be described as energy due to position relative to a standard position, normally chosen to be the surface of the earth. The potential energy of a body may be defined as the amount of work it can do when it moves from its actual position to the standard position chosen.

3.8 Formula for (gravitational) potential energy

Let a body be at rest on the surface of the earth. It is then raised to a vertical height h above the surface of the earth. The work required to do this will be force times distance and since the force is the weight mg, the work done on the body is mgh. Any variation of g with height is neglected. This work is stored as potential energy of the body and the body has the capacity to do this amount of work when it returns to the earth's surface. The potential energy is thus given by

Potential energy = mgh (zero at earth's surface)   [3.6]

Worked example 3.4
What is the potential energy (relative to the surface of the earth) of a 10 kg mass
(a) 100 m above the surface of the earth;
(b) at the bottom of a vertical mine shaft 1000 m deep.

Solution
Potential energy = mgh
(a) Potential energy = $10 \times 9.81 \times 100$ J
                      = 9.81×10^3 J
                      = 9.81 kJ

(b) Potential energy = $-10 \times 9.81 \times 1000$ J
                      = -9.81×10^4 J
                      = -98.1 kJ

The value is negative because the mass is below the surface of the earth.

3.9 Kinetic energy

A body may possess energy due to its motion as well as due to its position. For example, when a hammer is used to drive a nail, work is done on the nail by the hammer and hence the hammer must have possessed energy. Also a rotating flywheel possesses energy due to its motion. These are examples of the form of energy called kinetic energy. Kinetic energy may be described as energy due to motion. Only linear motion will be considered. The kinetic energy of a body may be defined as the amount of work it can do in being brought to rest.

3.10 Formula for kinetic energy

Let a body of mass m moving with speed v be brought to rest with uniform retardation by a constant force P in a distance s. Using [1.4]

$v^2 = u^2 + 2as$
$0 = v^2 - 2as$ since a is negative
or $s = v^2/2a$

Work done = force × distance
= Ps
= $Pv^2/2a$

However $P = ma$
and hence

Work done = $mav^2/2a$
= $\frac{1}{2} mv^2$

The kinetic energy is thus given by

Kinetic energy = $\frac{1}{2} mv^2$ [3.7]

3.11 Kinetic energy and work done

When a body with mass m has its speed increased from u to v in a distance s by a constant force P which produces acceleration a, then

$v^2 = u^2 + 2as$

and increase in kinetic energy is

$\frac{1}{2} mv^2 - \frac{1}{2} mu^2 = mas$
Thus since $P = ma$
increase in kinetic energy = Ps

However Ps = force × distance
= work done on body

Thus

Work done by forces acting on body = change of kinetic energy of body [3.8]

Equation [3.8] is sometimes called the work–energy theorem.

Worked example 3.5

A car of mass 1000 kg travelling at 30 m/s has its speed reduced to 10 m/s by a constant braking force over a distance of 75 m. Find the initial and final kinetic energies and the braking force.

Solution

Initial kinetic energy $= \frac{1}{2} mu^2$
$= 500 \times 30^2$
$= 4.5 \times 10^5$ J

Final kinetic energy $= \frac{1}{2} mv^2$
$= 500 \times 10^2$
$= 0.5 \times 10^5$ J

Change in kinetic energy $= 4.0 \times 10^5$ J

By [3.8]

Work done = change in kinetic energy
Hence Braking force $\times 75 = 4.0 \times 10^5$
or   Braking force $= 5333$ N

3.12 Conservation of energy

The principle of conservation of energy states that the total energy of a system remains constant. Energy cannot be created or destroyed but may be converted from one form to another.

If a crate slides from rest down a rough slope, then initially all its energy is potential. As it accelerates, some of its potential energy is converted into kinetic energy and some is used to overcome friction. This latter part is not lost but is converted into heat. At the bottom of the slope the energy will be purely kinetic assuming that the datum for potential energy is the bottom of the slope.

If a body falls freely under gravity and air resistance is negligible, then mechanical energy is conserved and loss of potential energy as the body falls equals the gain in kinetic energy as the speed increases.

If the motion involves friction or impact then the principle of conservation of mechanical energy cannot be used because some energy will be converted into heat and perhaps sound.

Worked example 3.6

A cyclist, with his bicycle, has mass 80 kg. He reaches the top of a hill, with slope 1 in 20 measured along the slope, at a speed of 2 m/s. He then free-wheels to the bottom of the hill where his speed has increased to 9 m/s. How much energy has been lost on the hill which is 100 m long?

Solution

If the hill is 100 m long and slopes at 1 in 20 along the slope, the cyclist has descended 100/20 = 5 m vertically. His potential energy loss is thus

$mgh = 80 \times 9.81 \times 5 = 3924$ J.

His increase in kinetic energy is

$\frac{1}{2} mv^2 - \frac{1}{2} mu^2 = 40 (81 - 4) = 3080$ J

By the principle of conservation of energy
Initial energy = final energy + loss of energy
Thus loss of energy (due to friction, etc)
= 3924 − 3080
= 844 J

3.13 Power

Power is the rate at which work is done or the rate at which energy is transferred, that is

$$\text{Power} = \frac{\text{work done}}{\text{time taken}} \qquad [3.9]$$

If work done is measured in joules and time in seconds, then power is in watts, (W). A power of 1 W means that work is being done at a rate of 1 J/s. Larger units of power are the kilowatt (kW) and the megawatt (MW) where 1 kW = 10^3 W and 1 MW = 10^6 W.

If work is being done by a machine moving at speed v against a constant force or resistance F, then, since work done is force times distance, work done per second will be Fv and hence the power in this case is

$$\text{Power} = Fv \qquad [3.10]$$

Worked example 3.7

A constant force of 2 kN pulls a crate along a level floor a distance of 10 m in 50 s. What power was used?

Solution

Work done = force × distance
= $2 \times 10^3 \times 10$
= 20×10^3 J

Power = work done/time taken
= $20 \times 10^3/50$
= 400 W

Worked example 3.8

A hoist operated by an electric motor has mass 500 kg. It raises a load of 300 kg vertically at a steady 0.2 m/s. Frictional resistance can be taken to be a constant 1200 N. Find the power required.

Solution
Total mass = 800 kg
Weight = 800 × 9.81
 = 7848 N

Total force = 7848 + 1200
 = 9048 N

Power = force × speed [3.10]
 = 9048 × 0.2
 = 1810 W
 = 1.81 kW

Worked example 3.9

The engine of a car has a power output of 42 kW. It can achieve a maximum speed of 120 km/h along the level. Find the resistance to motion. If the power output and resistance remained the same, what would be the maximum speed which the car could achieve up an incline of 1 in 40 along the slope if the car has mass 900 kg?

Solution
120 km/h = 120 × 10³/3600
 = 33.33 m/s

Power = resistance × speed
Resistance = 42 × 10³/33.33
 = 1260 N

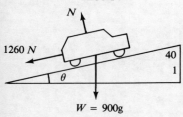
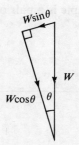

Fig. 3.4 Worked example 3.9

Total force down the incline
= frictional force + component of weight down incline
= 1260 + $mg \sin \theta$
= 1260 + 900 × 9.81/40
= 1260 + 221
= 1481 N

Power = force × speed
Maximum speed = 42000/1481
= 28.4 m/s
= 28.4 × 3600/10³
= 102 km/h

3.14 Moment, couple and torque

The moment of a force F about a point is its turning effect about the point and is the product of the force and the perpendicular distance from the point to the line of action of the force.

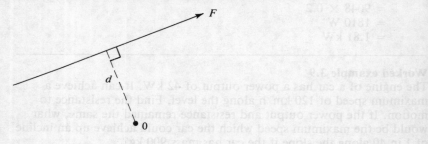

Fig. 3.5 Moment of a force

In Fig. 3.5 the moment of F about the point O is

Moment of a force = Fd [3.11]

A couple is a pair of equal and parallel but unlike forces as shown in Fig. 3.6.

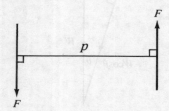

Fig. 3.6 Moment of a couple

It can easily be proved that the moment of a couple about any point in its plane is the product of one force and the perpendicular distance between them, that is

Moment of couple = Fp [3.12]

Examples of a couple include turning off a tap with finger and thumb and winding up a clock with a key.

The moment of a force or couple may be measured in newton metre (Nm).

Note that a couple may be balanced only by another equal and opposite couple acting in the same plane.

In engineering, the moment of a force or couple is called a torque. A spanner tightening a nut is said to exert a torque on the nut. Similarly, a belt passing round a pulley can exert a torque on it.

3.15 Work done by a constant torque

Let a force F turn a light rod OA with length r through an angle θ to OB as shown in Fig. 3.7.

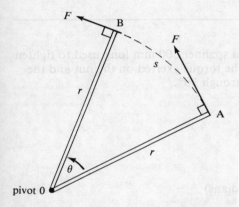

Fig. 3.7 Work done by a constant torque

The torque T_Q exerted about O is force times perpendicular distance from O

or $T_Q = Fr$ [3.13]

Now work done by F is F times distance moved. Hence

Work done $= Fs$

But s is the arc of a circle radius r. Hence

$s = r\theta$

where θ must be measured in radians.

Thus work done $= Fr\theta$
or work done $= T_Q\theta$ [3.14]

3.16 Power transmitted by a constant torque

If the rod in Fig. 3.7 rotates at n revolutions per second, then, in one second angle turned through $\theta = 2\pi n$ rad, and the work done per second will be $2\pi n T_Q$ by [3.14]. Power transmitted by a constant torque T_Q is thus

Power $= 2\pi n T_Q$
or Power $= \omega T_Q$ [3.15]

since angular speed ω is $2\pi n$.

The power will be in watts if n is in rev/s, ω is in rad/s and T_Q is in Nm.

Worked example 3.10
The force exerted on the end of a spanner 300 mm long used to tighten a nut is a constant 100 N. Find the torque exerted on the nut and the work done when the nut turns through 30°.

Solution
Torque $T_Q = Fr$
$= 100 \times 300 \times 10^{-3}$
$= 30$ Nm

Work done $= T_Q \theta$
$= 30 \times \pi/6$ (θ in radians)
$= 15.7$ J

Worked example 3.11
An electric motor is rated at 400 W. If its efficiency is 80%, find the maximum torque which it can exert when running at 2850 rev/min.

Solution
Power $= 2\pi n T_Q$
$n = 2850/60 = 47.5$ rev/s

Power $= 400 \times 0.8 = 320$ W

Torque $T_Q = 320/2\pi \times 47.5$
$= 1.07$ Nm

3.17 Work done by a variable torque

In practice, the torque is often variable. For example, the torque produced by a car engine is variable. Also, many machines require a varying torque for their operation. In such cases, the work done cannot be found using [3.14]. It can be found by a method similar to that used for a variable force.

Let the torque vary as in Fig. 3.8.

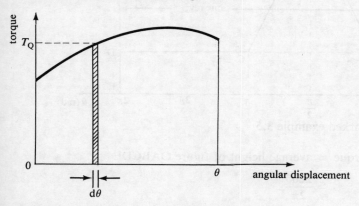

Fig. 3.8 Work done by a variable torque

The work done when the angular displacement is $d\theta$ is $T_Q d\theta$. This is the area of the shaded strip in Fig. 3.8. The total work done for the angular displacement θ is thus the area under the torque–angular displacement graph. For a variable torque then

Work done = area under torque–angular displacement graph   [3.16]

This formula should be compared with [3.2]. For non-linear graphs, the area may be found by any of the methods outlined in Sections 3.3, 3.4 and 3.5.

Worked example 3.12
A machine requires a variable torque as shown below.
Find the work done per revolution, the average torque over one revolution and the power required if the machine operates at 30 rev/min.

Solution
Work done = area under torque/θ graph
= area of triangle ABC + area of rectangle ADEO
= $\frac{1}{2}(\pi \times 600) + 2\pi \times 200$
= 700π
= 2200 J for one revolution

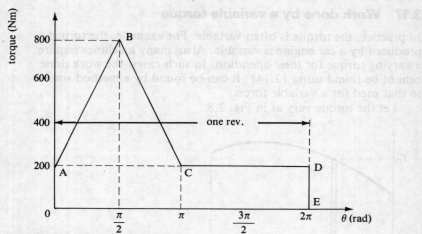

Fig. 3.9 Worked example 3.5

Average torque = average height of figure OABCDE

$$= \frac{2200}{2\pi}$$

$$= 350 \text{ Nm}$$

Power required = 2π (average torque)n
where n = 30/60 = 0.5 rev/s

Power = $2\pi \times 350 \times 0.5$
= 1100 W

Worked example 3.13
The runner of a turbine has mass 200 kg. It is fixed to a horizontal shaft of mass 100 kg and diameter 50 mm supported in two bearings at equal distances from the runner. The coefficient of friction between shaft and bearings is 0.01. Find

(a) the frictional force at each bearing;
(b) the frictional torque on the shaft;
(c) the power used to overcome friction in the bearings when the runner rotates at 2000 rev/min.

Solution

(a) Load m on each bearing is
½ (200 + 100) = 150 kg

Force exerted by shaft on each bearing is
mg = 150 × 9.81
= 1471 N

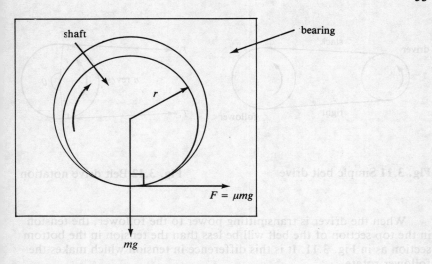

Fig. 3.10 Worked example 3.13

But frictional force $F = \mu$ (normal reaction)
$\qquad\qquad\qquad\qquad = \mu\, mg$

Thus frictional force on each bearing = 0.01×1471
$\qquad\qquad\qquad\qquad\qquad\qquad\quad = 14.7$ N

(b) Friction torque on shaft
$= Fr$ for each bearing
$= 2 \times 14.7 \times 50 \times 10^{-3}/2$ for both bearings
$= 0.735$ Nm

(c) Power used to overcome friction is, by [3.15],
$2\pi n T_Q = 2\pi \times 2000 \times 0.735/60$
$\qquad\quad = 154$ W

3.18 Power transmitted by belts

When power has to be transmitted from one shaft to another, then belt driving between pulleys on the shafts may be used, especially if the distance between the shafts is considerable.

Power is transmitted through friction between pulleys and belt. Neglecting slip, it is easy to verify that

$$\frac{\text{rev/s of driver}}{\text{rev/s of follower}} = \frac{\text{diameter of follower}}{\text{diameter of driver}} \qquad [3.17]$$

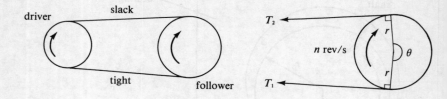

Fig. 3.11 Simple belt drive   **Fig. 3.12** Belt drive notation

When the driver is transmitting power to the follower, the tension in the top section of the belt will be less than the tension in the bottom section as in Fig. 3.11. It is this difference in tension which makes the follower rotate.

Clockwise torque on pulley = $T_1 r$ (see Fig. 3.12)
Anticlockwise torque on pulley = $T_2 r$

Since $T_1 > T_2$, the effective torque on the pulley is $(T_1 - T_2)r$
Since power transmitted by a constant torque is $2\pi n T_Q$

Power transmitted = $2\pi n(T_1 - T_2)r$ watts   [3.18]

In practice, T_1 may be two or three times T_2. The maximum possible value of $(T_1 - T_2)$ determines the maximum power which may be transmitted. Maximum possible $(T_1 - T_2)$ increases as the coefficient of friction between the belt and the pulley increases and as the angle of lap θ (see Fig. 3.12) increases. If $(T_1 - T_2)$ is increased too much, slipping will occur.

Worked example 3.14
The ratio of the tensions in the two sections of a driving belt is 2.8 to 1. The driver has diameter 500 mm. If the driver transmits 5 kW when it is rotating at 300 rev/min, find the tensions in the two sections of the belt.

Solution
Power transmitted = $2\pi n(T_1 - T_2)r$ from [3.18]

Thus $5 \times 10^3 = 2\pi \times \dfrac{300}{60} (T_1 - T_2) \times \dfrac{500 \times 10^{-3}}{2}$

or $5000 = 7.854 (T_1 - T_2)$
or $T_1 - T_2 = 637$ N   ... [1]

However if T_1 and T_2 are the tensions in the tight and slack sections of the belt the question states that

$T_1 = 2.8\, T_2$ ... [2]

Using [1] and [2] then gives

$2.8\, T_2 - T_2 = 637$
or $T_2 = 637/1.8 = 354$ N
and hence $T_1 = 991$ N

The required tensions are thus 991 N and 354 N.

Exercises

Objective type
Choose the ONE response which is the most appropriate.

3.1 The work done by the force P in Fig. 3.13 when it moves its point of application from A to B is
  A Pd
  B $Pd \sin \theta$
  C $Pd \cos \theta$
  D $Pd/\cos \theta$

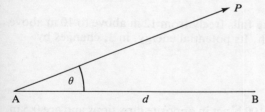

Fig. 3.13 Exercise 3.1

3.2 A body with weight 100 N is 10 m above the surface of the earth. Its gravitational potential energy, relative to the earth's surface is
  A 101.9 J
  B 1000 J
  C 9810 J
  D none of these

3.3 A 6 kg crate is pulled 5 m along a horizontal plane by a horizontal force of 10 N. The work done, in J, is
  A 30
  B 50
  C 60
  D 294

3.4 The unit used for measuring the moment of a couple could be the
  A kg m
  B kg rad
  C N/m
  D Nm

3.5 Which quantity is a vector?
  A Power
  B Kinetic energy
  C Work done
  D None of these

3.6 A unit used for measuring power could be the
  A N/ms
  B Nm
  C kg ms
  D kW

3.7 When a body falls freely, under gravity, from rest,
  A its potential energy decreases while its kinetic energy increases
  B its kinetic energy equals its potential energy
  C its kinetic energy decreases
  D its potential energy is proportional to the distance fallen

3.8 The kinetic energy of a body, with mass M, travelling at a given speed, is proportional to
  A M
  B $1/M$
  C $\sqrt{M}$
  D $1/\sqrt{M}$

3.9 A body with mass 10 kg falls freely from 12 m above to 10 m above the surface of the earth. Its potential energy, in J, changes by
  A 1177
  B 981
  C 196
  D none of these

3.10 Two parallel forces of 100 N act in opposite directions and are 0.5 m apart. The moment of the couple, in Nm, is
  A 25
  B 50
  C 100
  D 200

3.11 A train travels along a straight, level track at a constant 20 m/s and exerts a pull of 10 000 N. The power of the train, in kW, is
  A 200 000
  B 1962
  C 200
  D 20.4

3.12 The force acting on a body varies as in Fig. 3.14. The work done over the 4 km distance moved is

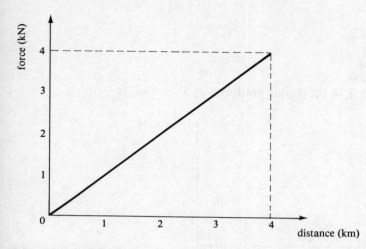

Fig. 3.14 Exercise 3.12

    A 16 MJ
    B 16 kJ
    C 8 MJ
    D 8 kJ

3.13 Which expression represents power?
    A force $\times$ displacement
    B force $\times$ velocity/time
    C force $\times$ velocity
    D mass $\times$ displacement/time

3.14 A winch raises a load of 10 kg vertically upwards. Figure 3.15 shows how the height of the load varies with time. The power consumed, in

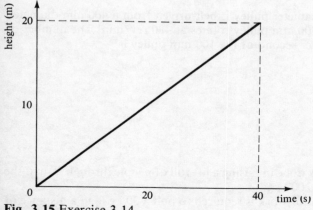

Fig. 3.15 Exercise 3.14

W, is
- A 5
- B 49
- C 196
- D 4900

3.15 In Fig. 3.16 the torque produced by *P*, in Nm, is

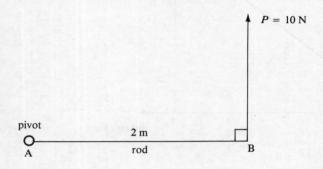

Fig. 3.16 Exercise 3.15

- A 5
- B 10
- C 20
- D none of these

3.16 The work done by the torque in Fig. 3.16 when the rod turns through 3 radians, in J, is
- A 6
- B 20
- C 30
- D 60

3.17 A 100 mm diameter pulley is belt-driven from a 400 mm diameter pulley. The 400 mm pulley rotates at 480 rev/min. The number of revolutions per second of the 100 mm pulley is
- A 2
- B 32
- C 120
- D 1920

Problems

3.18 Find the work done in raising a lift full of people through 20 m if the total mass is 2000 kg.

3.19 What amount of work is required to pull a 30 kg load a distance of 10 m up a smooth incline of 1 in 10 along the slope?

3.20 The force on a body varies with distance as shown below.

Force (kN)	0	7	14	25	31	28	17
Distance moved (m)	0	10	20	40	50	80	100

Find the work done as the body moves 100 m.

3.21 The force exerted on a truck varies with distance moved as shown in the table below.

Force (kN)	5.0	3.8	2.3	1.6	1.4	1.8	1.7	1.0	0
Distance (m)	0	10	20	30	40	50	60	70	80

Find the total work done during the 80 m.

3.22 What is the potential energy of a 20 kg mass (relative to the earth's surface) when it is
(a) 50 m above the surface of the earth;
(b) down a mine shaft 500 m deep;

3.23 A lift of mass 1000 kg moves from the second floor of a building 8 m above ground level to the sixth floor 24 m above ground level. What change in its potential energy has there been?

3.24 What is the kinetic energy of a mass of 0.08 kg travelling at 16 m/s?

3.25 Find the kinetic energy of a body of mass 2 kg when it falls, under gravity, a distance of 2 m from rest.

3.26 A body of weight 20 N falls freely, from rest, from a height of 10 m above the ground. Find the potential energy, the kinetic energy and the sum of the potential and kinetic energies
(a) just before the body is released;
(b) when it has fallen 2 m;
(c) just as it strikes the ground;
Comment on the answers obtained.

3.27 A man of mass 70 kg runs up a flight of 50 steps each 0.25 m high in 25 s. What is his rate of working?

3.28 A car engine has an output of 8 kW. Find the resistance to motion when the car travels at a steady speed of 15 m/s along a level road.

3.29 A 2 kg body is projected up a rough slope at 30° to the horizontal with an intial kinetic energy of 200 J. The coefficient of friction is 0.2. Find

(i) the retarding force,
(ii) the retardation,
(iii) the distance gone before coming to rest,
(iv) the gain in potential energy, and hence
(v) the loss of energy due to friction.

3.30 An electric water pump with overall efficiency 55 per cent raises 100 000 kg of water per hour through 30 m and delivers it at a speed of 1 m/s. Find the kinetic energy supplied to the water per second, the potential energy supplied to the water per second and the power expended in pumping the water. Also find the power input to the pump.

3.31 The engine of a car has a maximum power output of 50 kW. If the mass of the car is 1 t and the total resistance to motion is a constant 600 N, find the maximum speed which the car can achieve along the level and up an incline of 1 in 50 along the incline if the transmission has an efficiency of 70 per cent.

3.32 An engine with power 150 kW pulls a train with total mass 150 000 kg up an incline of 1 in 250 along the slope. The resistance to motion is a constant 20 N per 1000 kg of mass. Find
  (i) the total force down the incline
  (ii) the maximum uniform speed
which the train can achieve up the incline.

3.33 A mine-shaft 200 m deep, of cross-sectional area 12 m^2, has water in it 70 m deep. The water is pumped out to ground level in 6 hours and is delivered at 5 m/s. Find

  (i) the potential energy given to the water;
  (ii) the kinetic energy given to the water;
  (iii) the average power of the pump.

(Density of water is 1000 kg/m^3)

3.34 A turbine is driven by a waterfall where 50 m^3 of water per minute fall 10 m vertically. A power of 40 kW is developed. What fraction of the available energy is utilized? (1 m^3 of water has mass 1000 kg)

3.35 A torque of 30 Nm is applied to a shaft to turn it through 5 revolutions. What is the work done?

3.36 Find the torque at the shaft of an electric motor rated at 5 kW and with a speed of 1425 rev/min.

3.37 The torque required to operate a machine varies in the following way over two revolutions:
$0-\frac{1}{2}$ Revolution, torque increases linearly from zero to 800 Nm
$\frac{1}{2}-1$ Revolution, torque remains constant at 800 Nm
$1-3/2$ Revolution, torque drops abruptly to 500 Nm and then remains constant
$3/2-2$ Revolution, torque decreases linearly to zero.
Find the work done during a two revolution cycle, the average value of the torque and the power required by the machine if it operates at 240 rev/min.

3.38 A 200 mm pulley on an engine output shaft drives another pulley through a belt in which the ratio of tight to slack side tensions is 2.1 : 1. If the power output of the engine is 3 kW and its speed is 600 rev/min, find the tensions in the two sides of the belt.

3.39 A shaft exerts a force of 10 kN on a bearing. The coefficient of friction between shaft and bearing is 0.01. If the shaft diameter is 100 mm, find the power dissipated by friction when the rotation speed is 500 rev/min.

3.40 Find the maximum power which can be transmitted to a pulley with diameter 400 mm rotating at 180 rev/min by a belt when the angle of lap θ is 130°. The coefficient of friction μ between belt and pulley is

0.4. The maximum allowable belt tension is 500 N and the relationship between T_1 and T_2 is

$$2.3 \log \left(\frac{T_1}{T_2}\right) = \mu\theta$$

where θ is in radians.

3.41 The torque applied to a shaft varies as follows:

Torque (Nm)	35	100	70	45	25	10	0
Angular displacement	0	30π	60π	90π	120π	150π	180π

Draw a torque against angular displacement graph and find

(i) the total work done,
(ii) the average torque.

Chapter 4

Impulse and impact

Learning objectives

After reading this chapter and working through the exercises, you should be able to:
- define momentum and impulse;
- state the principle of conservation of (linear) momentum;
- solve problems involving change and conservation of momentum.

4.1 Impulse of a force

The impulse of a constant force P is defined as the product of the force and the time t for which it acts. The effect of this impulse on a body of mass m can be found using

$$v = u + at \qquad [4.1]$$

From [4.1] $at = v - u$
and hence $mat = m(v - u)$
Now $P = ma$
therefore $Pt = m(v - u)$ $\qquad [4.2]$

Since momentum is the product of mass and velocity, [4.2] can be stated as

Impulse of a constant force $= Pt =$ change in momentum produced. Impulse is a vector quantity and has the same units as momentum, namely kg m/s or Ns.

The impulse of a *variable* force P can be defined as

Impulse $= \int_0^t P \, dt$

where t is the time for which P acts.

Therefore Impulse $= \int_0^t m \dfrac{dv}{dt} \, dt$

using Newton's second law in the form

$P = m \dfrac{dv}{dt}$

or Impulse $= \int_u^v m \, dv = \Big[mv \Big]_u^v$

for constant mass

or Impulse $= m(v - u)$

where v and u are final and initial velocities respectively.

Hence, generally,

Impulse of a force $= \int_0^t P \, dt =$ change in momentum produced.

This relationship is sometimes referred to as the impulse-momentum theorem. It can be compared with the work-energy theorem stated as

Work done by a force = change in kinetic energy produced. [4.3]

4.2 Impulsive force

Suppose the force P is very large and acts for a very short time. During the time for which P acts, the body will move a very short distance which in many cases may be neglected. Under these conditions, the sole effect of the force can be measured by the impulse, or the change in momentum produced. In these cases the force is called an impulsive force. Theoretically, the force should be infinitely large and the time for which it acts infinitely small. These conditions are sometimes approached in practice, for example, in such situations as the collision of two snooker balls. Other examples where impulsive forces act include a hammer striking a nail and the impact of a bullet on a target.

Worked example 4.1
A nail of mass 0.02 kg is driven into a fixed wooden block. Its initial speed is 30 m/s and it is brought to rest in 5 ms. Find the impulse and the value of the force (assumed constant) on the nail.

Solution

Impulse = change in momentum of nail
= 0.02 × 30
= 0.6 Ns

Also   Impulse = Force × time

Therefore Force = $\dfrac{\text{Impulse}}{\text{Time}} = \dfrac{0.6}{0.005}$

= 120 N

Worked example 4.2

A football of mass 0.45 kg travels in a straight line along the ground and reaches a player at 10 m/s. He passes the ball along the ground to another player and in doing so alters the direction of motion of the ball by 90°. The ball leaves him at 8 m/s. Find the magnitude of the impulse given to the ball by the player.

Solution

Choose the directions Ox and Oy as shown in Fig. 4.1.

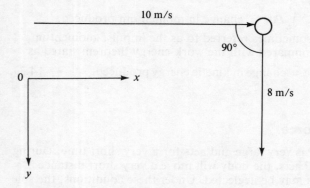

Fig. 4.1 Worked example 4.2

Initial velocity in the direction Ox is 10 m/s. Final velocity in the direction Ox is zero. Change in velocity in the direction Ox is −10 m/s. Initial velocity in the direction Oy is zero. Final velocity in the direction Oy is 8 m/s. Change in velocity in the direction Oy is 8 m/s. Thus the resultant change in velocity is

$\sqrt{(-10)^2 + 8^2} = \sqrt{164} = 12.8$ m/s

But   Impulse = change in momentum
Therefore Impulse = 0.45 × 12.8
= 5.76 Ns

4.3 Impact of water on a fixed surface

When a jet of water strikes a surface, there will be a change of momentum and hence there will be a force on the surface. When rain falls on the ground, there is a succession of small impulsive forces and in this and the previous situation, the force due to the water can be found by calculating the change in momentum per second.

Worked example 4.3
A jet of water of cross section 2000 mm^2 and speed 10 m/s strikes a fixed wall normally. Assuming that the speed of the water is reduced to zero by the impact, find the force exerted on the wall. Density of water is 1000 kg/m^3.

Solution
Volume of water striking the wall per second = $2000 \times 10^{-6} \times 10$ m^3

Since density = $\dfrac{\text{mass}}{\text{volume}}$

mass of water striking the wall per second

$= 2000 \times 10^{-6} \times 10 \times 10^3$ kg
$= 20$ kg

Therefore momentum destroyed per second

$= 20 \times 10$
$= 200$ Ns

Therefore   Force = rate of change of momentum
= 200 N

Worked example 4.4
Find the average pressure on the ground due to a rainfall of 10 mm in an hour. Assume that the velocity of a raindrop at the ground is that acquired after falling freely a distance of 10 m. The density of water is 1000 kg/m^3.

Solution
Velocity of raindrop can be found using
$v^2 = u^2 + 2as$
with $a = 9.81$ m/s^2 and $u = 0$, giving
$v = \sqrt{2 \times 9.81 \times 10}$
   $= 14.0$ m/s

Volume of rain falling in one second on an area of one square metre

$= \dfrac{10 \times 10^{-3} \times 1}{3600}$ m^3.

The mass of rain is thus $\dfrac{10 \times 10^{-3} \times 1 \times 10^3}{3600} = 2.78 \times 10^{-3}$ kg/m²

The momentum destroyed per second per square metre (assuming no rebound of drops)

$= 2.78 \times 10^{-3} \times 14.0$
$= 3.89 \times 10^{-2}$ Ns

and hence the average pressure on the ground due to the rain is 0.0389 N/m².

4.4 Impact of water on a moving surface

This is the principle underlying the operation of a water turbine such as the Pelton wheel. In this, a jet of water moving at high speed strikes a series of specially-shaped vanes closely spaced on the circumference of a rotating wheel.

For simplicity, assume that the vane turns the jet through 180° as shown in Fig. 4.2.

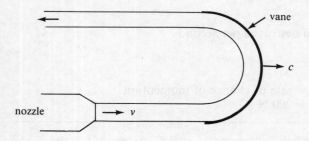

Fig. 4.2 Jet of water striking a moving vane

If the tangential speed of the vane is c, then the speed of the jet relative to the vane as it strikes the vane is $(v - c)$ and, neglecting friction, the speed of the jet leaving the vane and relative to the vane must also be $(v - c)$.

The change in speed relative to the vane is thus $2(v - c)$ and if m kg/s of water are intercepted by the wheel, the momentum change per second and hence the force on the wheel is $2m(v - c)$.

In practice, the force is less than this. This is mainly because there is friction and the angle turned through by the jet is less than 180°.

4.5 Conservation of linear momentum

Consider the direct collision of two spheres A and B as shown in Fig. 4.3.

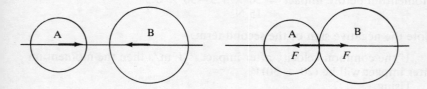

Fig. 4.3 Direct collision of two spheres

When they collide, then by Newton's third law, the force F exerted by A on B is equal and opposite to the force exerted by B on A. The time of contact is the same for both. The impulse of A on B is thus equal and opposite to the impulse of B on A. It then follows that the change in momentum of A is equal in magnitude to the change in momentum of B but is opposite in direction. The total change in momentum of this simple system produced by the collision is thus zero. This means that the momenta before and after the collision are equal. This is an example of the **principle of conservation of linear momentum** which may be stated as:

The total momentum of a system, in any direction, remains constant unless an external force acts on the system in that direction.

4.6 Impact of inelastic bodies

When two inelastic bodies collide, they remain together after the collision. The bodies show no tendency to return to their original shape if they are deformed by the collision. An example of an inelastic collision could be that of two railway trucks which become coupled on impact.

Problems of this type can be solved using the **principle of linear momentum** in the form

Momentum before impact = Momentum after impact
(in any direction) (in the same direction)

Although momentum is conserved, it is important to realize that energy is always lost in an inelastic collision in the form of heat, light, sound, etc.

Worked example 4.5
A railway wagon of mass 20 tonne travelling at 1.5 m/s collides with another of mass 30 tonne travelling in the opposite direction at

0.5 m/s. The wagons become coupled on impact. Find their common velocity after impact and the loss of kinetic energy.

Solution

Momentum before impact = $20 \times 1.5 - 30 \times 0.5$
$$= 15 \text{ Ns}$$

Note the negative sign of the second term.

If the common velocity after impact is V m/s then the momentum after impact will be $(20 + 30)V$.

Using

momentum before impact = momentum after impact
$$15 = (20 + 30)V$$
Therefore $\qquad V = 0.3$ m/s

in the original direction of the velocity of the 20 tonne wagon.

Kinetic energy = $\tfrac{1}{2}mv^2$

Therefore

Kinetic energy before impact = $\tfrac{1}{2}(20)(1.5)^2 + \tfrac{1}{2}(30)(0.5)^2$
$$= 26.25 \text{ J}$$

and

Kinetic energy after impact = $\tfrac{1}{2}(20 + 30)(0.3)^2$
$$= 2.25 \text{ J}$$

Therefore

Loss of kinetic energy = 24.0 J

Worked example 4.6

A piledriver of mass 2.5 tonne drives a pile of mass 500 kg vertically into the ground. The driver falls freely a vertical distance of 2 m before striking the pile and there is no rebound. Each blow of the driver drives the pile 0.2 m. Find the average value of the resistance of the ground to penetration.

Solution

The speed of the driver just before it strikes the pile can be found using

$v^2 = u^2 + 2as$
with $u = 0$
$\quad a = 9.81$ m/s^2
and $s = 2$ m

Therefore $v^2 = 2 \times 9.81 \times 2$
and $\qquad v = 6.26$ m/s

The momentum just before impact is thus $2.5 \times 10^3 \times 6.26$, that is, 15 650 Ns.

Since there is no rebound, the pile and driver have the same speed after impact and hence the momentum just after impact is $(2500 + 500)\,V$ where V is the common speed of pile and driver just after impact.

Using momentum before impact = momentum after impact
$$15\,650 = (2500 + 500)\,V$$
Therefore $\quad V = 5.22$ m/s.

The pile and driver are now brought to rest in 0.2 m by the ground resistance.

The retardation can be found using

$v^2 = u^2 + 2as$
with $u = 5.22$ m/s
$\quad\quad s = 0.2$ m
and $v = 0$

Therefore $0 = 5.22^2 + 2 \times 0.2 \times a$
and $\quad\quad a = -68.1$ m/s^2

The retarding force P can be found using

$P = ma$
$\quad = (2500 + 500)\,68.1$
$\quad = 204.3$ kN

Finally, the ground resistance, R, which is the sum of the weight of the pile plus driver and the retarding force, is

$R = 3000 \times 9.81 \times 10^{-3} + 204$
$\quad = 233$ kN.

The average ground resistance is thus 233 kN.

4.7 Impact of elastic bodies

In the previous section, dealing with inelastic collisions, the bodies involved were assumed to stay together after impact, that is, there was no rebound. An elastic body is one which tends to return to its original shape after impact. The compression produced by impact is followed by recovery or restitution of shape. When two elastic bodies collide, they rebound after the collision. An example of an elastic collision could be that between two snooker balls.

If the bodies are moving along the same straight line before impact, then the collision is called a direct collision. This is the only type of collision which will be considered in this text.

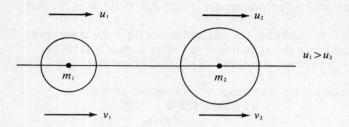

Fig. 4.4 Direct collision of two elastic spheres

Consider two elastic spheres colliding directly as shown in Fig. 4.4. By the principle of conservation of linear momentum

$$m_1 u_1 + m_2 u_2 = m_1 v_2 + m_2 v_2 \quad [4.3]$$

where u_1 and u_2 are the velocities just before impact and v_1 and v_2 the velocities just after impact. When the spheres are inelastic, v_1 and v_2 are equal and this equation is sufficient to solve problems where u_1 and u_2 are known. For elastic bodies, v_1 and v_2 will depend on the elasticity of the bodies. A measure of the elasticity is the coefficient of restitution e. For direct collisions this is defined as

$$e = -\left(\frac{v_1 - v_2}{u_1 - u_2}\right) \quad [4.4]$$

Note that all the velocities *must* be measured in the same direction as shown in Fig. 4.4. This equation is the result of experiments performed by Newton.

The values of e in practice vary between 0 and 1. If $e = 0$, the bodies are inelastic and $v_1 = v_2$. If $e = 1$, the bodies are perfectly elastic. Only in this latter case will no energy be lost in the collision.

Equations [4.3] and [4.4] can now be used to solve problems on the direct collision of elastic spheres.

Worked example 4.7
A body of mass 2 kg moving with speed 5 m/s collides directly with another of mass 3 kg moving in the same direction at 4 m/s. The coefficient of restitution is 2/3. Find the velocities after the collision.

Solution
By the principle of conservation of linear momentum
$m_1 u_1 + m_2 u_2 = m_1 v_1 + m_2 v_2$
or $2 \times 5 + 3 \times 4 = 2v_1 + 3v_2$
$\qquad\qquad 22 = 2v_1 + 3v_2 \qquad \ldots [1]$

By the definition of e

$$e = -\left(\frac{v_1 - v_2}{u_1 - u_2}\right)$$

or $\dfrac{2}{3} = -\left(\dfrac{v_1 - v_2}{5 - 4}\right)$

$-2 = 3v_1 - 3v_2$ ... [2]

Adding [1] and [2] gives

$20 = 5v_1$

Therefore $v_1 = 4$ m/s

and hence by [1]

$22 = 8 + 3v_2$

Therefore $v_2 = \dfrac{14}{3}$ m/s

Therefore the velocities are 4 m/s and 14/3 m/s respectively in the original directions.

Worked example 4.8
A railway wagon has a mass of 15 tonne and is moving at 1.0 m/s. It collides with a second wagon of mass 20 tonne moving in the opposite direction at 0.5 m/s. After the collision, the second wagon has its speed changed to 0.4 m/s and its direction of travel reversed.

Find the velocity of the 15 tonne wagon after the collision, the coefficient of restitution and the loss of kinetic energy.

Solution
By the principle of conservation of linear momentum

$$m_1 u_1 + m_2 u_2 = m_1 v_1 + m_2 v_2$$

or $15\,000 \times 1.0 - 20\,000 \times 0.5 = 15\,000 \times v_1 + 20\,000 \times 0.4$

Note that the second term on the left hand side of the equation has a negative sign.

This gives

$15 v_1 = -3$

Therefore $v_1 = -0.2$ m/s
This means that after the collision the 15 tonne wagon has had its direction of travel reversed and its speed changed to 0.2 m/s.

Coefficient of restitution $e = -\left(\dfrac{v_1 - v_2}{u_1 - u_2}\right)$

$= \dfrac{(-0.2) - 0.4}{1.0 - (-0.5)}$

$= 0.4$

Note the signs in the expression for e.

Kinetic energy before impact $= \tfrac{1}{2}(15\,000)(1.0)^2 + \tfrac{1}{2}(20\,000)(0.5)^2$
$= 10\,000$ J

Kinetic energy after impact $= \tfrac{1}{2}(15\,000)(0.2)^2 + \tfrac{1}{2}(20\,000)(0.4)^2$
$= 1900$ J

Therefore loss of kinetic energy $= 8100$ J

Exercises

Objective type
Choose the ONE response which is the most appropriate.

4.1 A suitable unit for measuring impulse could be the
  A kg m s
  B kg/ms
  C N/s
  D Ns

4.2 When two inelastic bodies collide
  A momentum is conserved but kinetic energy is not conserved
  B momentum and kinetic energy are both conserved
  C neither momentum nor kinetic energy is conserved
  D momentum is not conserved but kinetic energy is conserved

4.3 When a jet of water strikes a rigid vertical wall, the force exerted by the water on the wall can be calculated by
  A using the principle of conservation of momentum
  B using the principle of conservation of energy
  C equating force to rate of change of momentum
  D finding the change in kinetic energy

4.4 Which of the following quantities is NOT a vector quantity?
  A Impulse
  B Coefficient of restitution
  C Weight
  D Momentum

4.5 Coefficient of restitution must have magnitude
  A greater than unity
  B between 0.5 and unity
  C between zero and 0.5
  D between zero and unity

4.6 A hammer of mass 1 kg strikes a nail at 5 m/s and rebounds at 1 m/s.
The change in the momentum of the hammer in Ns, is
A 6
B 5
C 4
D 1

4.7 Two trucks, each of mass 5 tonne, collide head-on. Just before the impact their speeds are 2 m/s and 1 m/s. If they become coupled together on impact, their common speed just after impact, in m/s, is
A 3
B 1.5
C 1
D 0.5

Problems

4.8 A nail of mass 0.01 kg travelling horizontally at 10 m/s as it enters a fixed wooden block is brought to rest in 1 ms. Find the average resistance of the wood.

4.9 A railway wagon of mass 5 tonne strikes a pair of buffers at 2 m/s and rebounds at 1.5 m/s. The time of contact with the buffers is 0.4 s. What is the average force exerted by the buffers?

4.10 Water issues from a hose at 600 kg/min. The jet strikes a vertical wall at 20 m/s. Assuming no rebound of the water, estimate the force on the wall.

4.11 Waves strike a vertical sea-wall at 20 m/s. Estimate the pressure on the wall due to the destruction of momentum if 1 m^3 of sea water has mass 1025 kg.

4.12 In a Pelton-wheel turbine, the jet of water emerges from a nozzle with area 10^{-3} m^2 with a speed of 180 m/s. The tangential speed of the vanes is half the speed of the jet. Find (a) the mass of water striking the wheel per second and (b) the force on the wheel. Assume that the vanes turn the jet through 180° and ignore all losses. Density of water is 1000 kg/m^3

4.13 A 5 kg mass falls freely from a height of 2 m onto a fixed horizontal surface. It rebounds to a height of 0.5 m. The time of impact is 0.025 s. Find the change in momentum and the average force between the mass and the surface.

4.14 A pile of mass 0.5 tonne is driven vertically into the ground by a pile-driver of mass 1.2 tonne. If the driver falls freely 4 m before striking the pile and the ground resistance is assumed to be constant at 150 kN, find how far the pile is driven into the ground if there is no rebound.

4.15 A body of mass 1 kg moving at 3.5 m/s collides with another body of mass 2.5 kg moving in the same direction along the same line at 1 m/s. If the coefficient of restitution is 0.8, find the velocities of the bodies after impact.

4.16 (a) What is the impulse of a force?
(b) A hammer of mass 1 kg, moving horizontally at 5 m/s, strikes a nail of mass 0.025 kg and drives it 25 mm into a vertical timber plank. Assuming that there is no rebound, calculate:

    (i) the common speed after impact
    (ii) the time of motion of the nail
    (iii) the average resistance of the timber
    (iv) the loss of kinetic energy.

4.17 (a) State the principle of conservation of linear momentum.
(b) A railway wagon of mass 8 tonne rolls from rest down an incline 1 km long and falling 1 m in 100 m along the incline. The resistance to motion is a constant 500 N. Find the speed of the wagon at the bottom of the incline. When it reaches the bottom of the incline, the wagon collides inelastically with a second wagon of mass 10 tonne and both move off together horizontally after the collision. Find the common speed just after the impact.

4.18 (a) Explain the meaning of the term coefficient of restitution.
(b) A railway wagon of mass 20 tonne, moving at 3 m/s, collides with a second wagon of mass 15 tonne and at rest. The coefficient of restitution is 0.6. Find the loss of kinetic energy during the impact.

Part 3

Mechanics of Fluids

Chapter 5

Fluids at rest

Learning objectives

After reading this chapter and working through the exercises you should be able to:
- state that fluid pressure at a point acts equally in all directions;
- explain that pressure increases with depth below its surface and determine total force and centre of pressure for plane submerged surfaces;
- relate the above to civil and structural engineering problems.

5.1 Fluids

A fluid may be either a liquid or a gas. Fluids differ from solids in their behaviour when subjected to a shearing stress. A solid will return to its original state once the shearing stress is removed, provided that the elastic limit is not exceeded. A fluid is permanently deformed by any shear stress however small, that is, it flows under the action of a shearing stress. It follows that in a fluid at rest there can be no shear stress and hence all fluid forces on surfaces either bounding or submerged in a fluid must act at right angles to those surfaces. Cold tar and pitch might at first sight appear to be solids. However, if stressed sufficiently long, they will flow and hence are fluids.

Liquids and gases behave quite differently when pressure or temperature changes. The volume of a gas can change considerably when its pressure or temperature changes whereas the volume of a liquid changes very little. A gas is thus relatively compressible whereas a liquid is relatively incompressible.

5.2 Molecular structure of fluids

In a solid, the molecules are close together and the forces between them are strong. The molecules are not free to move and can only vibrate about a mean position.

Molecules in a fluid are free to move around and the forces between them are much less than in a solid. However, in a liquid the forces are strong enough to produce a liquid surface when the liquid is in a container. A gas on the other hand always occupies all of its container.

In fluid mechanics the molecular structure of fluids may usually be ignored. Instead, they are treated as if they were continuous. This means that the average density, pressure, etc, are constant or that they vary continuously with distance or with time.

5.3 Fluid properties

Several important fluid properties are considered below.

Density

This is the mass of unit volume of the fluid. It is often represented by the symbol ρ (rho) and is measured in kg/m³. The density of water is usually taken to be 1000 kg/m³ although the value decreases slowly as temperature increases and increases very slowly as pressure increases.

Specific weight

This is the weight of unit volume of the fluid and hence equals ρg where g is the acceleration due to gravity. It is often represented by the symbol w or γ and is measured in N/m³. If ρ is 1000 kg/m³, then w (or γ) will be 9.81×10^3 N/m³.

Specific gravity (S.G.)

This is also known as the relative density. It is the ratio of the density of a substance to the density of some reference substance, usually water at 4°C (277 K). Since specific gravity is a ratio it has no units. The density of mercury is 13 600 kg/m³ and if the density of water is 1000 kg/m³ then the specific gravity of mercury is 13.6.

Viscosity

Shear forces cannot exist in a fluid at rest. However, shear forces do exist in a moving fluid. The motion of the coffee in a cup produced by stirring eventually ceases due to shear forces in the liquid. Water spilled on a horizontal surface spreads rapidly compared with tar or thick oil. The property which determines the rate of spreading is called the viscosity. It is a measure of the resistance of a fluid to shear stress.

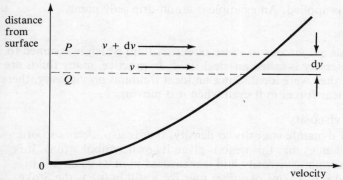

Fig. 5.1 Velocity distribution near a plane surface

Consider a fluid flowing in thin layers, all parallel to each other and to a plane surface – see Fig. 5.1. The layer of fluid in contact with the surface will be kept at rest by surface irregularities. The shear force F between any two layers P and Q of area A, separated by a small distance dy and having a small velocity difference of dv is directly proportional to A and dv and inversely proportional to dy.

That is

$$F = \mu A \frac{dv}{dy} \qquad [5.1]$$

where μ (mu) is a constant for the liquid. The constant is called the dynamic (or absolute) viscosity of the fluid.

Equation [5.1] may be written as

$$\mu = \frac{F/A}{dv/dy} \qquad [5.2]$$

The ratio F/A is the shear stress and dv/dy is the rate of shear strain or velocity gradient. Equation [5.2] is called Newton's law of viscosity. If F is measured in N, A in m², dv in m/s and dy in m, then μ will be in Ns/m². Another unit for μ still in use is the poise (P); 1 Poise = 0.1 Ns/m². For water at 20°C (293 K) μ is 1.00×10^{-3} Ns/m². As temperature increases, the viscosity of gases increases and the viscosity of liquids decreases.

A fluid, such as water, which obeys [5.2] and which has a constant value of viscosity at a constant temperature is called a Newtonian fluid. Non-Newtonian fluids include plastic fluids and thixotropic fluids. Plastic fluids do not flow until a certain minimum shear stress is exceeded. When flow does occur, the relationship between F/A and dv/dy may or may not be linear. If it is linear, the plastic fluid is then known as a Bingham plastic. An example is sewage sludge. A thixotropic fluid has a viscosity which decreases with the time for which the shear stress is applied. An example is a non-drip jelly paint.

Ideal fluid

All real fluids are viscous to some extent. A fluid which is assumed to have zero viscosity is called an ideal fluid. In practice, many fluids are treated as if they were ideal. Since an ideal fluid has no viscosity, there can be no shear forces in it even when it is moving.

Kinematic viscosity

The ratio of dynamic viscosity to density, that is μ/ρ, occurs so often in fluid mechanics that this ratio is given its own symbol ν (nu). It is called the kinematic viscosity and is measured in m^2/s when μ is in Ns/m^2 and ρ is in kg/m^3. Another unit for ν still in use is the Stoke (St). 1 Stoke = 10^{-4} m^2/s. At 20°C (293 K) the kinematic viscosity of water is 1.01×10^{-6} m^2/s.

Surface tension

The surface of a liquid is, in some ways, like an elastic membrane in a state of tension. If a line is imagined to be drawn in the liquid surface, then the force per unit length acting perpendicular to one side of this line and in the plane of the surface is called the surface tension. It is a phenomenon associated with the forces between the molecules at the interface between two fluids which do not mix. Surface tension is measured in N/m. For pure water at 20°C (293 K) the surface tension is 7.4×10^{-2} N/m. This value may be greatly reduced by the addition of a detergent such as washing-up liquid. Surface tension is the property of a liquid which causes raindrops to have the smallest possible surface area consistent with their volume. This makes them practically spherical. Surface tension also produces capillary rise (or fall) of liquids in narrow pores, for example, the rise of water in concrete.

The height h to which a liquid with density ρ will rise or fall in a vertical narrow tube with diameter d is given by

$$h = \frac{4S \cos \theta}{\rho g d} \qquad [5.3]$$

where S is the surface tension of the liquid and θ is the angle of contact. This is the angle, shown in Fig. 5.2, between the tube wall and the tangent to the liquid meniscus, measured through the liquid. For clean water and clean glass $\theta = 0°$ and for mercury and glass θ is approximately 135° (in this case $\cos \theta$ is negative and the liquid level in the tube is below that outside it).

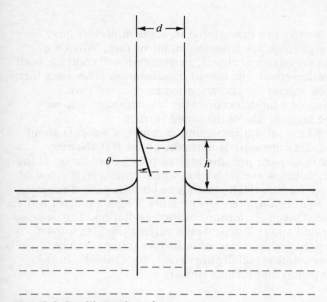

Fig. 5.2 Capillary elevation

Surface tension forces are often small enough to be ignored in fluid mechanics. However, they may have to be included in situations such as the flow in small hydraulic models.

Compressibility

The effect of pressure on the volume of a fluid can be expressed in terms of its compressibility which is defined as

$$\text{Compressibility} = \frac{\text{change in volume}}{\text{original volume} \times \text{pressure change}} \qquad [5.4]$$

The reciprocal of compressibility is widely used. It is called the bulk modulus of elasticity K and

$$K = \frac{\text{original volume} \times \text{pressure change}}{\text{change in volume}} \qquad [5.5]$$

The units for K are N/m^2.

Water is relatively incompressible and the density changes produced by compression can usually be ignored. However, when dealing with a phenomenon such as water hammer where a large increase of pressure can be produced by a valve closing suddenly compressibility must be considered. The bulk modulus of water is about 2.2×10^3 kN/m². For steel, the value is about 1.7×10^5 kN/m² indicating that water is about 80 times more compressible than steel.

Vapour pressure

All liquids evaporate to some extent, that is, some molecules have enough kinetic energy to escape from the liquid surface. When the space above the liquid surface is closed, evaporation will continue until the rate of evaporation equals the rate of condensation. The space then contains a saturated vapour. Evaporation increases rapidly with temperature. Boiling of a liquid occurs when its saturation vapour pressure equals the pressure above the liquid surface.

At 20°C (293 K) the saturation vapour pressure of water is about 2.3 kN/m². Atmospheric pressure is normally about 100 kN/m².

When a liquid flows, low pressure regions may be produced. If the pressure in these falls below the saturation vapour pressure, a cloud of vapour bubbles may form in the liquid. These bubbles can be swept along by the flow to a region where the pressure is high enough for the bubbles to collapse. This phenomenon is called cavitation. The collapse can be extremely rapid and if it occurs on a surface the high impact forces produced by the collapse can cause surface pitting or even structural failure. Cavitation is especially important, for example, in the design of pump impellers and turbine runners.

5.4 Pressure at a point in a fluid

Since there can be no shear force in a fluid at rest, the fluid forces on a surface immersed in or bounding a fluid at rest must act at right angles to the surface.

The average pressure intensity p on an element dA of plane surface with area A immersed in a fluid is given by

$$p = \frac{dF}{dA} \qquad [5.6]$$

where dF is the normal force due to the fluid. If the pressure intensity is uniform over the area A then

$$p = \frac{F}{A} \qquad [5.7]$$

where F is the total normal force due to the fluid.

If F is measured in N and A is in m², then p is in N/m² or Pascal (Pa). The pressure intensity (normally known simply as the pressure) at a point in a fluid is the value of p given by equation 5.7 when A becomes infinitely small.

Worked example 5.1
The pressure at the bottom of a rectangular water tank with a horizon-

tal base measuring 2 m × 2.5 m is 25 kN/m². Find the total force on the base due to the water.

Solution

Pressure = $\dfrac{\text{Force}}{\text{Area}}$

Thus Force = Pressure × Area
 = 25 × 10³ × 2 × 2.5
 = 125 × 10³ N
 = 125 kN

5.5 Variation of pressure in a static fluid

(a) Horizontally
Consider an imaginary cylindrical volume of water as shown in Fig. 5.3

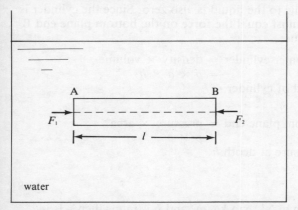

Fig. 5.3 Notation for the variation of pressure on a horizontal line in a liquid

and let the axis of the cylinder be horizontal. The horizontal forces on the cylinder are F_1 and F_2 acting on the plane ends A and B of the cylinder. If the cylinder is at rest F_1 equals F_2. Further, if the ends of the cylinder are made infinitely small they become points and the pressures p_1 and p_2 at these two points must also be equal. It follows that the pressures at any two points on the same horizontal level in a continuous mass of fluid at rest are equal.

(b) Vertically
Consider now an imaginary cylinder of water as shown in Fig. 5.4 and let the axis of the cylinder be vertical. The resultant force on the curved

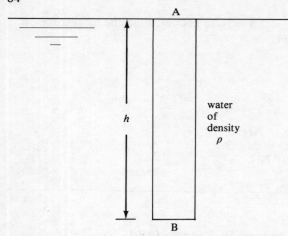

Fig. 5.4 Notation for the variation of pressure with depth in a liquid

sides of the cylinder due to the surrounding water is zero. The force on the top plane end A due to the liquid is also zero. Since the cylinder is at rest, its weight mg must equal the force on the bottom plane end B due to the water pressure.

Now mass m of water cylinder = density × volume
$$= \rho \times a \times h$$
where a = area of end of cylinder
Thus $mg = g\rho a h$

Also   force on bottom plane end = pressure × area
$$= p \times a$$
where p = water pressure at depth h

Thus $pa = g\rho a h$
or $p = g\rho h$ [5.8]

If g is measured in m/s^2, ρ in kg/m^3 and h is in m, then p is in N/m^2 (or Pa).

Since the density of the water is assumed to be constant, [5.8] applies only to incompressible fluids. Equation [5.8] gives the pressure at depth h below the free surface of a liquid with density ρ. Note that pressure due to the liquid is directly proportional to depth h.

If the pressure at the liquid surface is p_0 then the pressure at depth h in the liquid would be $p_0 + g\rho h$. Pressure p_0 could be atmospheric pressure p_{ATM}.

5.6 Equality of pressures in all directions at a point

Consider a sphere immersed in a liquid as shown in Fig. 5.5.

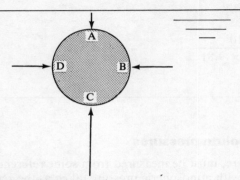

Fig. 5.5 Pressure at a point

The pressure p_A at the highest point A will be less than the pressure p_C at the lowest point C. At B and D, the pressures p_B and p_D will be equal and will have magnitude greater than p_A but less than p_C. If the radius of the sphere becomes infinitely small, the values of p_A, p_B, p_C and p_D will all become equal in magnitude. This shows that, at a point in a fluid, the fluid pressure acts equally in all directions.

5.7 Pressure head

Since pressure p is proportional to depth h it is possible and often convenient to measure pressure not as a force per unit area but as a height or head h of fluid where h is given by [5.8] rearranged to give

$$h = \frac{p}{\rho g} \qquad [5.9]$$

The pressure head h will be in metres if p is measured in N/m², ρ in kg/m³ and g in m/s².

Worked example 5.2
Express a pressure of 105 kN/m² in (a) metres of water and in (b) metres of oil with specific gravity 0.88.

Solution
(a) $p = g\rho h$

Thus $h = \dfrac{105 \times 10^3}{10^3 \times 9.81}$

if the density of water is taken to be 10^3 kg/m³. That is

$h = 10.7$ m of water.

(b) $p = g\rho h$

Thus $h = \dfrac{105 \times 10^3}{10^3 \times 0.88 \times 9.81}$

or $h = 12.2$ m of oil

5.8 Gauge and absolute pressures

Pressure, like temperature, must be measured from some reference value. If it is measured with atmospheric pressure taken as the zero, then the pressure is known as the gauge pressure. If a perfect vacuum is chosen as the zero, then the pressure is known as the absolute pressure. Many pressure measuring devices, for example, a tyre pressure gauge, record the difference between atmospheric pressure and the pressure being measured. They therefore record gauge pressure. Equation [5.8] gives the gauge pressure at depth h in a liquid since the value of pressure given does not include atmospheric pressure.

Gauge pressure may be positive or negative since pressure may be above or below atmosphere but absolute pressure is always positive.

The relationship between absolute, gauge and atmospheric pressures is

Absolute pressure = Gauge pressure + Atmospheric pressure [5.10]

This relationship is illustrated in Fig. 5.6.

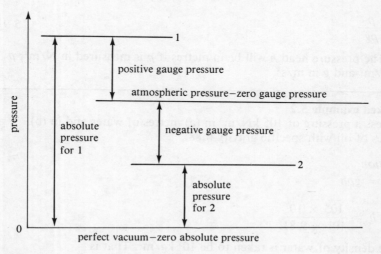

Fig. 5.6 Absolute and guage pressures

Worked example 5.3
Find the gauge and absolute pressures 20 m below the surface of a lake when the atmospheric pressure is 100 kN/m².

Solution
Gauge pressure = $g\rho h$
= $9.81 \times 10^3 \times 20$
= 196×10^3 N/m²
= 196 kN/m²

Absolute pressure = 196 + 100
= 296 kN/m²

5.9 The mercury barometer

To find the absolute pressure, one requires a knowledge of the atmospheric pressure p_{ATM}. This can be found using the mercury barometer as illustrated in Fig. 5.7.

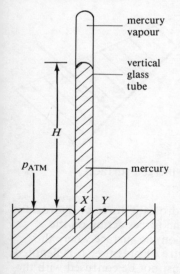

Fig. 5.7 A mercury barometer

Since points X and Y are on the same horizontal level
$p_X = p_Y$
However $p_Y = p_{\text{ATM}}$
and $p_X = g\rho_{\text{Hg}}H + p_{\text{vap}}$

where p_{vap} is the saturated vapour pressure of mercury. Now p_{vap} is only of the order of 1 mm of mercury while H is of the order of 750 mm of mercury. Thus p_{vap} may be neglected.

Hence $p_{ATM} = g\rho_{Hg}H$ [5.11]

The mean value of the atmospheric pressure at sea level is called standard atmospheric pressure. It is the pressure when H is 760 mm of mercury with specific gravity 13.59 and equals 101.3 kN/m² if g is taken to be 9.81 m/s². A pressure of 100 kN/m² is called a bar and 10^{-3} bar, the millibar (mb), is a pressure unit used by meteorologists.

5.10 Centre of pressure and centroid

Hydrostatic forces act at right angles to submerged or retaining surfaces. Also, the hydrostatic pressure increases linearly with depth below the free liquid surface. The distribution of pressure over one side of two plane surfaces is shown in Fig. 5.8. One surface is vertical and one is inclined.

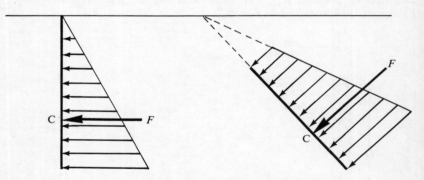

Fig. 5.8 Pressure distribution over two submerged surfaces

In Fig. 5.8, F denotes the resultant hydrostatic force or hydrostatic thrust on the surface, that is, the sum of the forces acting on the elements of area of the surface. The point on the surface where the hydrostatic thrust acts is known as the centre of pressure. It is denoted by C in Fig. 5.8. The centre of pressure must not be confused with the centroid of the surface. The centroid of an area is a geometric property. It may be defined as follows. The centre of gravity of a sheet of material of uniform thickness is the point through which its weight may be supposed to act. If the thickness of the sheet is made smaller, the position of the centre of gravity does not change. When the sheet is infinitely thin, it has area but no weight and the centre of gravity is then known as the centroid of the area. The position of the centroid depends on the shape of the area. The centroid will be denoted by the letter G.

5.11 Hydrostatic thrust and position of centre of pressure for a rectangular plane surface

Consider first the case where the submerged surface is vertical with one edge lying in the free liquid surface. The pressure variation over one side of the surface is linear, from zero at the liquid surface S to $\rho g l$ at the lowest point T as shown in Fig. 5.9. The average pressure is $\frac{1}{2}\rho g l$.

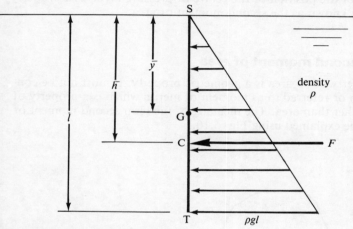

Fig. 5.9 Vertical submerged surface

If the area of the surface is A, then F is given by

F = Resultant force = average pressure × area
$= \frac{1}{2} \rho g l A$

Now if the length ST is l, then the depth $\bar{y}$ of the centroid G below the free liquid surface is $\bar{y} = \frac{1}{2}l$ and hence

$F = A g \rho \bar{y}$

Since the pressure diagram is triangular, F will act through the centroid of the triangle and hence C lies a vertical distance of $\frac{1}{3}l$ above T. The vertical depth $\bar{h}$ of the centre of pressure is thus

$\bar{h} = \frac{2}{3}l$

In this case G lies above C. This will be found to be always the case except when ST is horizontal when G and C are coincident.

If the surface ST is inclined at an angle α to the horizontal, it may easily be shown that

$F = Ag\rho\bar{y}$

where $\bar{y}$, the vertical depth of G, is now $\frac{1}{2}l \sin \alpha$ and that

$\bar{h} = \frac{2}{3}l \sin \alpha$

In more complex situations, for example, when the surface is not rectangular or when a rectangular surface does not have one side lying in the liquid surface, the determination of the position of the centre of pressure is more difficult. However it is possible to derive formulae, one for the hydrostatic thrust and one for the position of the centre of pressure which can be used for any surface inclined at any angle. The formula for the position of the centre of pressure involves a property of an area known as the second moment of area.

5.12 Second moment of area

This property of an area is a geometric property. It must not be confused with or referred to as moment of inertia which is a property of matter rather than area. The meaning of the term second moment of area can be explained using Fig. 5.10.

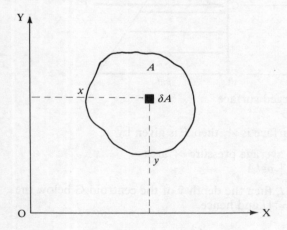

Fig. 5.10 Notation for second moment of area

If δA is an element of a plane area A, then the first moment of area of δA about OY is $x \, \delta A$. The second moment of area of δA about OY is $x (x \, \delta A)$, that is, $x^2 \, \delta A$. The second moment of area of the whole area A about OY is $\Sigma x^2 \, \delta A$. This can be denoted by the symbol I_{OY}. Similarly, I_{OX} will be $\Sigma y^2 \, \delta A$. The units for second moment of areas may be m⁴, mm⁴, etc.

5.13 Formulae for hydrostatic thrust and position of centre of pressure – plane immersed surface

Consider first the case where the surface is vertical. Let the plane of the immersed surface intersect the liquid surface in a line MM. Denote the depths of the centroid G and the centre of pressure C by $\bar{y}$ and $\bar{h}$ respectively – see Fig. 5.11. Consider forces on one side of the surface only.

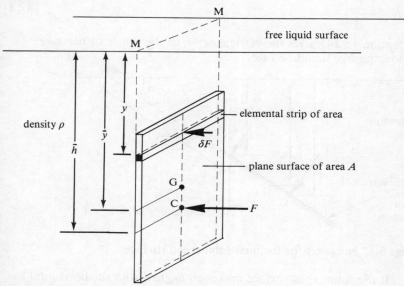

Fig. 5.11 Notation for vertical submerged surface

The force δF on an element of area δA is

δF = pressure × area
    = $\rho g y \, \delta A$ [5.12]

where y = depth of the element below the liquid surface.
For the whole area A, the total force F is

$F = \Sigma \delta F$
  = $g\rho \, \Sigma y \, \delta A$ since g and ρ are constants.

However $\Sigma y \, \delta A$ is the first moment of area of A about MM and this is also $A\bar{y}$.

Thus $F = A g \rho \bar{y}$ [5.13]

Note that the hydrostatic thrust F is the product of the area A and the pressure $\rho g \bar{y}$ at the centroid of the area.

To find the position of the centre of pressure C, take moments about MM.

Moment of δF about MM is $y\, \delta F$. For the whole area A, the total force moment is $\Sigma y\, \delta F$ or $\rho g\, \Sigma y^2\, \delta A$ using equation [5.12].

Now $\Sigma y^2\, \delta A$ is the second moment of area I_{MM} of A about MM. The total force moment about MM is also $F\bar{h}$.

Thus $F\bar{h} = \rho g\, I_{MM}$
or $A\, g\rho \bar{y}\, \bar{h} = \rho g\, I_{MM}$ using [5.13];

or $\bar{h} = \dfrac{I_{MM}}{A\bar{y}}$ [5.14]

Equation [5.14] gives the vertical depth of the centre of pressure below the free liquid surface.

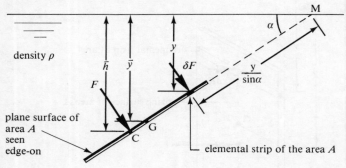

Fig. 5.12 Notation for inclined submerged surface

If the submerged surface makes an angle α with the horizontal as shown in Fig. 5.12 then it may be shown that

$F = Ag\rho \bar{y}$ as before [5.13]

and $\bar{h} = \dfrac{I_{MM} \sin^2 \alpha}{A\bar{y}}$ [5.15]

5.14 Parallel axis theorem

The value to be used for I_{MM} in [5.14] and [5.15] clearly depends on the position of the submerged surface relative to the line MM. However, the value of the second moment of area I_{CG} of any particular shape about an axis through its centroid is a constant. The value of I_{CG} may be modified to give I_{MM} by using an equation known as the parallel axis theroem. This may be stated as

$I_{MM} = I_{CG} + \dfrac{A\bar{y}^2}{\sin^2 \alpha}$ [5.16]

In this formula, I_{CG} is the second moment of area of the area A about a line through its centroid and *parallel to the line* MM.

The area, position of centroid and value of I_{CG} are given in Table 5.1 for three common shapes.

Table 5.1 Properties of common shapes

Shape	Area	Centroid position	Second moment of area I_{CG} about axis shown
rectangle	lb	$y_G = b/2$	$I_{CG} = lb^3/12$
triangle	$hb/2$	$y_G = h/3$	$I_{CG} = bh^3/36$
circle	$\pi d^2/4$	$y_G = d/2$	$I_{CG} = \pi d^4/64$

Worked example 5.4
A gate in a vertical dam wall is 2 m high by 3 m wide. The top edge of the gate is 5 m below the free water surface. Find the hydrostatic thrust on the gate and the position of the centre of pressure (Fig. 5.13).

Solution
Thrust $F = A g \rho \bar{y}$
$A = 3 \times 2 = 6 \text{ m}^2$
$\bar{y} = 5 + 1 = 6 \text{ m}$

Thus $F = 6 \times 9.81 \times 10^3 \times 6$ since the density of water is 10^3 kg/m^3
$\quad\quad = 353 \times 10^3 \text{ N}$
$\quad\quad = 353 \text{ kN}$

The hydrostatic thrust is thus 353 kN.

$I_{MM} = I_{CG} + A\bar{y}^2$

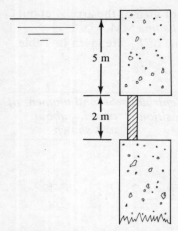

Fig. 5.13 Worked example 5.4

$$I_{CG} = \frac{lb^3}{12} = \frac{3 \times 2^3}{12}$$

$$= 2 \text{ m}^4$$

Note: l and b must be carefully identified.

Thus $I_{MM} = 2 + 6 \times 6^2$
$\qquad\quad = 218 \text{ m}^4$

$$\bar{h} = \frac{I_{MM}}{A\bar{y}} = \frac{218}{6 \times 6}$$

$$= 6.06 \text{ m}$$

The centre of pressure thus lies 6.06 m below the water surface.

Worked example 5.5
A circular gate in a reservoir wall has diameter 1.5 m. Its lowest point is 2.0 m below the water surface and the highest point 1.0 m – see Fig. 5.14. Find the hydrostatic thrust on the gate and the position of the centre of pressure.

Solution
$F = Ag\rho\bar{y}$

$$A = \frac{\pi d^2}{4} = \frac{\pi (1.5)^2}{4}$$

$$= 1.77 \text{ m}^2$$

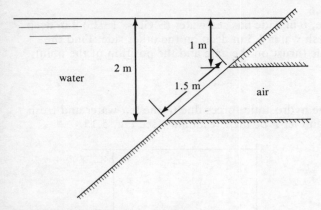

Fig. 5.14 Worked example 5.5

$\bar{y} = \frac{1}{2}(1 + 2)$
  $= 1.5$ m

Thus $F = 1.77 \times 9.81 \times 10^3 \times 1.5$
  $= 26.0 \times 10^3$ N
  $= 26.0$ kN

The hydrostatic thrust is thus 26.0 kN.

$I_{MM} = I_{CG} + \dfrac{A\bar{y}^2}{\sin^2\alpha}$

$I_{CG} = \dfrac{\pi d^4}{64} = \dfrac{\pi(1.5)^4}{64}$

  $= 0.25$ m^4

$\sin \alpha = 1/1.5 = 2/3$

Thus $I_{MM} = 0.25 + \dfrac{1.77 \times 1.5^2}{(2/3)^2}$

  $= 9.21$ m^4

and $\bar{h} = \dfrac{I_{MM}\sin^2\alpha}{A\bar{y}} = \dfrac{9.21 \times (2/3)^2}{1.77 \times 1.5}$

  $= 1.54$ m

The centre of pressure thus lies 1.54 m below the water level in the reservoir.

Worked example 5.6

A vertical lock-gate, 6 m wide has sea-water (S.G. = 1.025) 9 m deep on one side and fresh water 4.5 m deep on the other side. Find the resultant hydrostatic thrust on the gate and the position of the point where it acts.

Solution

Let F_1 and F_2 be the hydrostatic forces due to the sea-water and fresh water respectively and let F be their resultant – see Fig. 5.15.

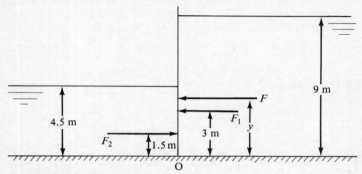

Fig. 5.15 Worked example 5.6

$F_1 = A_1 \, g\rho\bar{y}_1$ and acts $1/3 \, (9) = 3$ m above O
$F_2 = A_2 \, g\rho\bar{y}_2$ and acts $1/3 \, (4.5) = 1.5$ m above O

$A_1 = 6 \times 9 = 54 \text{ m}^2$
$A_2 = 6 \times 4.5 = 27 \text{ m}^2$

$\bar{y}_1 = \tfrac{1}{2} (9) = 4.5$ m
$\bar{y}_2 = \tfrac{1}{2} (4.5) = 2.25$ m

Thus $F_1 = 54 \times 9.81 \times 10^3 \times 1.025 \times 4.5$
$= 2443 \times 10^3$ N
$= 2443$ kN

and $F_2 = 27 \times 9.81 \times 10^3 \times 2.25$
$= 596 \times 10^3$ N
$= 596$ kN

Thus resultant thrust $F = F_1 - F_2$
$= 2443 - 596$
$= 1847$ kN

To find where F acts, take moments about O.
This gives $F_1 \times 3 - F_2 \times 1.5 = Fy$
where y = distance above O to the point where F acts.

Thus 2443 × 3 − 596 × 1.5 = 1847 y
or 1847 y = 6435
or y = 3.48 m

Resultant thrust is thus 1847 kN acting at a point 3.48 m above the bottom of the gate.

Worked example 5.7

A masonry dam has cross-section as shown in Fig. 5.16. The depth of water behind its vertical side is 3.6 m. Find the force per m run of the dam due to the water, the position where it acts and the minimum coefficient of friction required to prevent the dam sliding. The density of masonry is 2300 kg/m³.

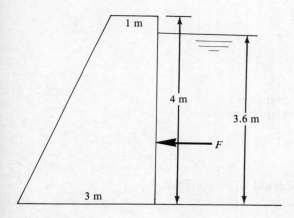

Fig. 5.16 Worked example 5.7

Solution
Water force $F = Ag\rho\bar{y}$
Consider one metre length of dam
Then $F = (3.6 \times 1) \times 9.81 \times 10^3 \times (3.6/2)$
   $= 63.6 \times 10^3$ N/m
   $= 63.6$ kN/m

F acts 1/3 (3.6) = 1.2 m above the base of the dam.
Weight of one metre length of dam is

W = cross sectional area × density × g
   $= (4 \times 1 + \frac{1}{2}(4 \times 2)) \times 2300 \times 9.81$
   $= 181 \times 10^3$ N
   $= 181$ kN

Minimum coefficient of friction required $= \mu \dfrac{F}{W} = \dfrac{63.6}{181} = 0.35$

Exercises

Objective type
Choose the ONE response which is the most appropriate.

5.1 If a liquid has density 880 kg/m³ its specific gravity is
   A 88 000
   B 880
   C 88
   D 0.88

5.2 An ideal fluid
   A is incompressible
   B has zero viscosity
   C has very low density
   D has zero surface tension

5.3 Dynamic viscosity could be measured in
   A Nsm²
   B N/sm²
   C Ns/m²
   D Nm/s²

5.4 Kinematic viscosity could be measured in
   A poise
   B Ns/m²
   C s/m²
   D m²/s

5.5 A tank is in the form of a cube with side 3 m. Its base is horizontal and it contains oil with S.G. 0.9 to a depth of 1 m. The force on the base, due to the oil, in kN, is
   A 79 500
   B 238
   C 88.3
   D 79.5

5.6 Which of the following sets of pressures are equivalent (S.G. of mercury is 13.6),

	kN/m²	m of water	mm of mercury
A	50	0.51	375
B	100	10.2	750
C	50	10.2	375
D	100	5.1	750

5.7 The gauge pressure in a water pipe is 60 kN/m². If atmospheric pressure is 760 mm of mercury, the absolute pressure, in kN/m², is
  A 41.4
  B 101.4
  C 161.4
  D none of these

5.8 A gauge on a pump inlet pipe records a negative gauge pressure of 20 kN/m². If the atmospheric pressure is 1.05 bar, what is the absolute pressure in the pipe, in kN/m²?
  A 20
  B 85
  C 105
  D 135

5.9 The magnitude of the hydrostatic force on one side of a submerged surface is the product of
  A area, liquid density and depth of centroid
  B area, g, liquid density and depth of centroid
  C area, g, liquid density and depth of centre of pressure
  D none of these

5.10 The centre of pressure of a vertical submerged surface
  A always lies below the centroid of the surface
  B always lies above the centroid
  C coincides with the centroid
  D may lie above or below the centroid

5.11 The second moment of area of a circular plate with diameter 2 m about a diameter, in m⁴, is
  A 4π
  B $\pi/4$
  C $\pi/16$
  D $\pi/64$

5.12 A vertical gate is 3 m square and has its upper edge horizontal and lying in a free water surface. The hydrostatic thrust on one side of the gate, in kN, is
  A 132 000
  B 265
  C 177
  D 132

5.13 A vertical surface 1 m square has its upper edge horizontal and 1.5 m below a free water surface. The hydrostatic thrust on one side of the surface, in kN, is
  A 13.2
  B 17.7
  C 19.6
  D 24.5

Problems

5.14 What head of water is equivalent to a pressure of five standard atmospheres? (S.G. of mercury is 13.6)

5.15 Find the pressure, in kN/m^2, at the bottom of a reservoir 45 m deep.

5.16 A pressure gauge on a water pipe indicates a negative gauge pressure of 220 mm mercury. Atmospheric pressure is $102\ kN/m^2$. What is the absolute pressure in the water pipe
(a) in kN/m^2;
(b) in mm of mercury?
(S.G. of mercury is 13.6)

5.17 If the pressure at the centre of an anticyclone is 1050 mb, what is the pressure in mm of mercury?

5.18 A tank open to the atmosphere contains 2.6 m of water covered by 1.8 m of oil with S.G. 0.91. What is the pressure at the bottom of the tank?

5.19 A rectangular water tank has a vertical end 8 m wide. The water depth in the tank is 2 m. Find the magnitude and position of the thrust on the end due to the water.

5.20 Just after it is poured, concrete behaves as a liquid with S.G. 2.4. Find the force per metre run of form used to pour a wall 4 m high.

5.21 A vertical circular flood gate of diameter 1.5 m is pivoted at its highest point. It has water on one side to a depth of 4 m above its highest point. Find the hydrostatic thrust on the gate and the moment of the force, about the highest point of the gate, just sufficient to open the gate against the thrust of the water.

5.22 A vertical sluice gate 2.8 m wide has the water 3.3 m above the bottom on one side and 2.4 m above the bottom on the other side. Find the magnitude and location of the resultant thrust on the gate.

5.23 A dam wall at 60° to the horizontal has an outlet closed by a circular plate flush with the wall and of diameter 1 m. The water depth above the centre of the plate is 4 m. Find the force on the plate due to the water and the position of the centre of pressure.

5.24 Find the force due to the water pressure, acting on the pivot of the gate shown in Fig. 5.17 if the gate is 3 m wide.

5.25 A vertical gate across a channel is 7 m wide. There is sea water (S.G. 1.02) 8.7 m deep on one side and fresh water 3.6 m deep on the other side. Find the resultant thrust on the gate and the position of the centre of pressure. What would the resultant thrust be if the depth on both sides was 8.7 m?

5.26 A circular plate 1.5 m in diameter closes the outlet to a dam as shown in Fig. 5.18. Find the hydrostatic thrust on the plate and find the value of the force P required to just keep the gate closed.

5.27 A vertical wall 5 m high is subject to a uniform wind pressure of 0.92 kN/m^2. Find the overturning moment about the base of the wall per metre run of wall.

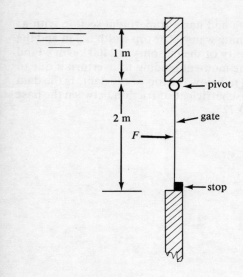

Fig. 5.17 Problem 5.24

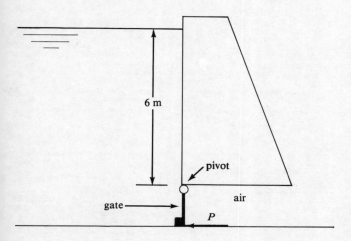

Fig. 5.18 Problem 5.26

5.28 A masonry dam is 50 m long and has a trapezoidal section with a vertical face 15 m high retaining water. The top width is 3 m and the base width is 13 m. The density of the masonry is 2400 kg/m^3. Find the force on the dam and the moment tending to overturn it due to the water when the level is 3 m below the top of the dam. Is the dam safe against sliding when the coefficient of friction between the base and the soil underneath is 0.7?

Chapter 6

Fluids in motion

Learning objectives
After reading this chapter and working through the exercises, you should be able to:
- define steady flow, uniform flow and discharge velocity;
- state the three energy components of a fluid in motion;
- derive Bernoulli's equation (including energy losses by friction) and explain its limitations in use;
- apply Darcy's formula ($h_f = \lambda l v^2 / 2gd$) to pipe sizing problems;
- apply Chezy's formula ($v = C\sqrt{mi}$) to problems of uniform flow in channels.

6.1 Steady flow and uniform flow

Conditions in a flowing fluid may change with time or with position within the flow. It is convenient to distinguish between different types of flow. If, at a particular point in a fluid, conditions such as flow velocity and pressure do not change with time then the flow is said to be steady. An example of steady flow is the flow at a point in a pipe conveying water with a constant speed. Problems involving steady flow are much easier to analyse than those where the flow is unsteady, that is, where conditions such as flow velocity and pressure at a point change with time. Unsteady flow occurs in a pipe while a valve is being closed or opened, thereby causing velocity and pressure to change.

If at all points in a flow the velocities at a particular instant are equal, then flow in that region is said to be uniform. An example of uniform flow could be the flow of water in a straight pipe with constant cross-sectional area. In non-uniform flow, at any instant, the velocity varies from point to point. Non-uniform flow occurs in a tapering section of a water pipe.

There are four possible combinations of steady or unsteady flow with uniform or non-uniform flow. Steady and uniform flow occurs in a pipe when the flow is at a constant rate in a straight pipe section of constant cross-sectional area. If the flow is at a constant rate through a tapering pipe section, then the flow is steady and non-uniform. An example of unsteady and uniform flow occurs in a pipe section of constant cross-sectional area when a pump in the pipeline is switched on causing acceleration of the flow. If this acceleration occurs in a tapering pipe section, the flow is unsteady and non-uniform.

6.2 Laminar flow and turbulent flow

In laminar flow (also known as streamline flow) the fluid may be considered to move in a series of infinitesimally thin layers all sliding over each other. The fluid particles in the layers have no velocity at right angles to the direction of movement of the layers, that is, there is no mixing of the fluid. Laminar flow is dominated by viscous shear forces and is governed by Newton's law of viscosity—equation [5.2]. Laminar flow is most likely to occur in situations involving viscous liquids moving slowly in narrow channels. It may be analysed mathematically.

In turbulent flow, the fluid particles move along erratic paths and there is movement and hence energy transfer at right angles to the flow direction, that is, there is mixing of the fluid. Turbulent flow is most likely to occur in situations involving liquids with low viscosity moving with high velocity in wide channels. Turbulent flow cannot yet be analysed exactly mathematically and experimentally derived formulae are used in calculations on turbulent flow.

6.3 Streamlines

A streamline is an imaginary, continuous line drawn in a flow at a particular instant so that at that instant there is no fluid flow across the line. The velocity of every fluid particle on the streamline is represented by a tangent to the streamline. The definition of a streamline means that streamlines cannot join up or intersect and that free liquid surfaces and bounding surfaces must be composed of streamlines. A series of streamlines drawn in a flow is known as a flow pattern. Flow patterns are useful for describing flows. For example, in uniform flow, the flow pattern must consist of a series of straight parallel lines. If they are not straight, velocity changes from point to point on the streamline. If the lines are not parallel, flow velocity must be increasing or decreasing. Converging streamlines indicate accelerating flow and diverging streamlines, decelerating flow – see Fig. 6.1

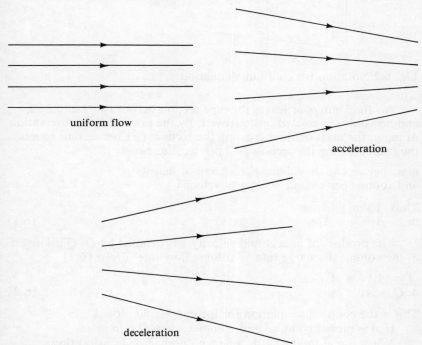

Fig. 6.1 Flow patterns

Flow patterns may be investigated experimentally, in the case of liquids by injecting a dye into the liquid flow and in the case of a gas by injecting smoke into the gas flow. When flow is laminar, the dye or smoke appears as a thin line known as a streakline. If the flow is turbulent, the line breaks up and the dye or smoke diffuses through the fluid.

6.4 Continuity equation

Consider an ideal incompressible fluid flowing through a tapering pipe section as shown in Fig. 6.2. Let the cross-sectional area and flow velocity be A_1 and v_1 respectively, at the section 1–1 and A_2 and v_2 at the section 2–2.

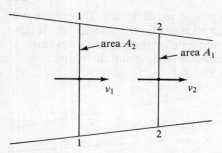

Fig. 6.2 Notation for continuity equation

No fluid enters or leaves the pipe section between 1–1 and 2–2 and no voids are created or destroyed. By the principle of conservation of mass, the mass of fluid crossing the section 1–1 per second equals the mass crossing the section 2–2 per second. Now

mass per second = volume per second × density ρ
and volume per second = area × velocity

Thus $A_1 v_1 \rho = A_2 v_2 \rho$
or $\quad A_1 v_1 = A_2 v_2$ [6.1]

The product of area A and velocity v is denoted by Q. Q is known as the volume discharge rate or volume flow rate. From [6.1]

$Q = A_1 v_1 = A_2 v_2$
or $Q = Av$ [6.2]

This is the continuity equation for incompressible flow.

If A is measured in m^2 and v in m/s, then Q is in m^3/s.

When a real fluid (that is, one with non-zero viscosity) flows through a pipe, the velocity varies across a section of the pipe. In fact it is zero at the pipe walls. Equation [6.2] can still be used, however, in the form

$Q = A\bar{v}$ [6.3]

where $\bar{v}$ is now the mean flow velocity.

Worked example 6.1

A water pipe has diameter 200 mm and the flow rate is 0.04 m^3/s. What is the mean flow velocity?

Solution

$Q = A\bar{v}$

where $A = \dfrac{\pi d^2}{4}$

$$= \frac{\pi \times 0.2^2}{4}$$

$$= 0.0314 \text{ m}^2$$

Thus $\bar{v} = \dfrac{Q}{A}$

$$= \frac{0.04}{0.0314}$$

$$= 1.27 \text{ m/s}$$

Worked example 6.2
A 250 mm diameter pipe A branches into two pipes B and C. Pipe B has diameter 100 mm and pipe C has diameter 200 mm. The mean flow velocity in B is 0.25 m/s and the flow rate in C is 3 litre/s. Find the mean flow velocity in pipe C and the flow rate and mean flow velocity in pipe A.

Solution
For pipe C $\quad Q_C = A_C \bar{v}_C$
Since $\quad$ 1 litre/s = 10^{-3} m^3/s
$Q_C = 3 \times 10^{-3}$ m^3/s

Thus $3 \times 10^{-3} = \left(\dfrac{\pi \times 0.2^2}{4} \right) \bar{v}_c$

and hence $\bar{v}_C = \dfrac{4 \times 3 \times 10^{-3}}{\pi \times 0.2^2}$

or $\bar{v}_C = 0.0955$ m/s

Since flow rate in A = flow rate in B + flow rate in C
$Q_A = Q_B + Q_C$
$Q_B = A_B \bar{v}_B$

$$= \left(\frac{\pi \times 0.1^2}{4} \right) \times 0.25$$

$= 1.96 \times 10^{-3}$ m^3/s

Thus $Q_A = 3 \times 10^{-3} + 1.96 \times 10^{-3}$
or $\quad Q_A = 4.96 \times 10^{-3}$ m^3/s
However $Q_A = A_A \bar{v}_A$

Thus $\bar{v}_A = \dfrac{4.96 \times 10^{-3}}{(\pi/4)(0.25)^2}$

or $\bar{v}_A = 0.101$ m/s

6.5 The Bernoulli equation

A fluid may possess energy and hence can do work by virtue of its position, its velocity or its pressure.

Consider a large water tank as shown in Fig. 6.3, with the water at a constant level.

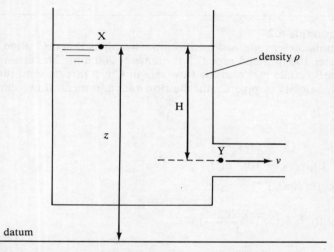

Fig. 6.3 Notation for potential and velocity heads

A mass M of water at X has potential energy Mgz referred to the datum. The height z, which has the dimensions of length, is called the **potential head**.

If the mass M eventually flows out of the outlet with velocity v, it will have kinetic energy $\tfrac{1}{2}mv^2$ where

$$\tfrac{1}{2}mv^2 = mgH$$

From this equation

$$H = \dfrac{v^2}{2g}$$

The kinetic energy may thus be rewritten as $mg(v^2/2g)$. The quantity

$v^2/2g$, which has the dimensions of length, is called the **kinetic (or velocity) head**.

If the outlet pipe has a close fitting, frictionless piston as shown in Fig. 6.4, then each second the piston would move a distance v. The work done per second would be pAv since work done is force times distance and force is pressure times area. Now Av is the volume flowing out per second.

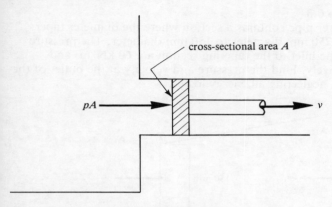

Fig. 6.4 Notation for pressure head

Thus

work done per second = p (volume per second)
= p (mass per second/density)

For mass M, then, the work done and hence the energy possessed by mass M, is pM/ρ. This may be rewritten as $Mg(p/\rho g)$. The quantity $p/\rho g$ is called the **pressure head**. It may easily be shown to have the dimensions of length.

The total energy E of a mass M of water is thus

$$E = Mgz + Mg\left(\frac{p}{\rho g}\right) + Mg\left(\frac{v^2}{2g}\right)$$

or $\dfrac{E}{Mg} = z + \dfrac{p}{\rho g} + \dfrac{v^2}{2g}$ [6.4]

The quantity E/Mg is the energy per unit weight. It is the sum of the potential, velocity and pressure heads. It is called the **total energy head** and it could be measured in metres of liquid. By the principle of conservation of energy, the total energy of a system remains constant.

Thus $z + \dfrac{p}{\rho g} + \dfrac{v^2}{2g}$ = a constant [6.5] for an ideal fluid. Equation [6.5] is called the Bernoulli equation. It was first formulated in 1738 by the Swiss mathematician Daniel Bernoulli. The equation is probably the best known equation in fluid mechanics. Strictly speaking, it applies only to the steady flow of an ideal, incompressible fluid along a streamline but it may be used with sufficient accuracy in many practical situations.

Worked example 6.3
A horizontal water pipe contains a section where the diameter tapers uniformly from 250 mm diameter to 150 mm diameter. The pressure and velocity at the inlet to the tapering section are 60 kN/m² and 0.8 m/s, respectively. Find the pressure and velocity at the outlet of the tapering section, neglecting all energy losses.

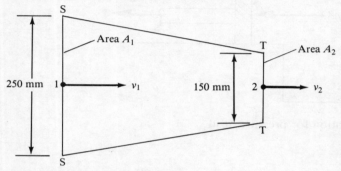

Fig. 6.5 Worked example 6.3

Solution
$Q = A_1 v_1 = A_2 v_2$

Thus $\dfrac{\pi}{4} (0.25)^2 \, 0.8 = \dfrac{\pi}{4} (0.15)^2 \, v_2$

and $v_2 = (0.25/0.15)^2 \, 0.8$
$= 2.22$ m/s

Applying the Bernoulli equation to the two points 1 and 2 on the pipe axis at sections SS and TT gives

$$z_1 + \frac{p_1}{\rho g} + \frac{v_1^2}{2g} = z_2 + \frac{p_2}{\rho g} + \frac{v_2^2}{2g}$$

However the pipe is horizontal and hence $z_1 = z_2$.

Thus $\dfrac{60 \times 10^3}{10^3 g} + \dfrac{0.8^2}{2g} = \dfrac{p_2}{10^3 g} + \dfrac{2.22^2}{2g}$

or $\quad 60 + 0.32 = \dfrac{p_2}{10^3} + 2.46$

Thus $p_2 = 57.9 \times 10^3 \text{ N/m}^2$
$= 57.9 \text{ kN/m}^2$

Pressure and velocity at the outlet are thus 57.9 kN/m² and 2.22 m/s respectively.

6.6 Loss of pressure head in pipes

Loss of pressure head in a pipe can be due to several factors. These include losses due to fluid viscosity, roughness of the pipe walls, changes in cross-sectional area of the pipe, pipe bends, valves and flow meters.

If the Bernoulli equation [6.5] is applied to two points on the axis of a pipe (inlet subscript 1 and outlet subscript 2) then the equation is written as

$$z_1 + \dfrac{p_1}{\rho g} + \dfrac{v_1^2}{2g} = z_2 + \dfrac{p_2}{\rho g} + \dfrac{v_2^2}{2g} + \text{head loss}$$

where the head loss term is the sum of the pressure head losses due to the factors mentioned above.

6.7 Reynolds number for pipe flow

The criterion used to determine whether flow in a pipe is laminar or turbulent is the value of a dimensionless set of quantities known as the Reynolds number. For a circular pipe of diamter d in which a fluid with density ρ and dynamic viscosity μ flows with mean velocity v, the Reynolds number Re is given by

$$Re = \dfrac{\rho d v}{\mu} \quad\quad\quad [6.6]$$

If $Re < 2000$ the flow will be laminar and if $Re > 4000$ the flow will be turbulent. If $2000 < Re < 4000$ flow may be laminar or turbulent depending on the conditions prevailing. Note from [6.6] that turbulent flow is most likely to occur when density, diameter and velocity are large and viscosity is low.

6.8 The Darcy–Weisbach formula

This equation gives the pressure head loss due to fluid viscosity and

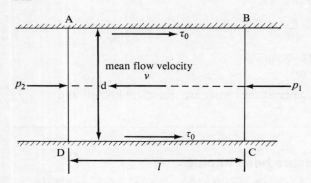

Fig. 6.6 Notation for Darcy – Weisbach formula

pipe roughness in a pipe in which the flow is incompressible, steady, uniform and turbulent. Consider a cylinder ABCD of fluid with length l and diameter d as shown in Fig. 6.6

Since there is no acceleration, the force on the cylinder due to the pressure difference, that is, $(p_1 - p_2)\pi d^2/4$ must balance the force due to the shear stress at the pipe walls. If the shear stress at the pipe walls is τ_0 the shear force is $\tau_0 l \pi d$.

Thus $(p_1 - p_2)\pi d^2/4 = \tau_0 l \pi d$.

or $\tau_0 = \dfrac{(p_1 - p_2)\, d}{4l}$ [6.7]

Experiment indicates that

$\tau_0 = kv^2$ [6.8]

where k is a constant. Using [6.6] and [6.7]

$$\dfrac{(p_1 - p_2)\, d}{4l} = kv^2$$ [6.9]

However pressure head loss h_f is just $(p_1 - p_2)/\rho g$ where ρ is the fluid density.
Thus, from [6.9],

$$\dfrac{p_1 - p_2}{\rho g} = \dfrac{4lkv^2}{d\rho g}$$

or $h_f = \dfrac{8k}{\rho}\, \dfrac{l}{d}\, \dfrac{v^2}{2g}$ If $8k/\rho$ is replaced by λ, then

$h_f = \dfrac{\lambda l}{d}\, \dfrac{v^2}{2g}$ [6.10]

This equation is the Darcy–Weisbach formula for circular pipes. The coefficient λ is a friction coefficient or friction factor. It is dimensionless and its value depends on either Reynonds number Re or on the pipe roughness or on both depending on flow conditions and on the pipe material.

In some textbooks, λ in [6.10] is replaced by f and in others by $4f$. Only the form of the equation as in [6.10] will be used in this text.

Worked example 6.4
A water pipe of diameter 150 mm runs full bore and the mean flow velocity is 0.4 m/s. If the friction coefficient is 0.016, find the head loss due to friction in a 1 km length of pipe.

Solution

$$h_f = \frac{\lambda l}{d} \frac{v^2}{2g}$$

$$= \frac{0.016 \times 1000 \times 0.4^2}{0.15 \times 2 \times 9.81}$$

$$= 0.87 \text{ m}$$

Worked example 6.5
Estimate the diameter of a horizontal water pipe, 300 m long, suitable for carrying 300 litre/s when running full bore if the pressure head drop is not to exceed 5 m. The friction coefficient is 0.02. Consider only friction losses and assume that the flow is turbulent.

Solution

$$h_f = \frac{\lambda l}{d} \frac{v^2}{2g}$$

and $Q = Av = \pi d^2 v/4$

Thus $h_f = \dfrac{\lambda l}{2dg} \left(\dfrac{4Q}{\pi d^2}\right)^2$

$$= \frac{8\lambda l Q^2}{\pi^2 g d^5}$$

and $5 = \dfrac{8 \times 0.02 \times 300 \times 0.3^2}{\pi^2 \times 9.81 \times d^5}$

Thus $d^5 = 8.92 \times 10^{-3}$
and $d = 0.389$ m

The required diameter is thus 389 mm.

Worked example 6.6

The decrease in elevation between the inlet and outlet of a 300 mm diameter water supply pipe 400 m long in 5 m. If the pressure at the inlet is 200 kN/m² and the flow rate is 0.1 m³/s, find the pressure head loss due to friction and the pressure at the outlet. All other losses, apart from friction, are negligible and the friction coefficient for the pipe material is 0.018. What is the Reynolds number for the flow? The viscosity of water is 1.14×10^{-3} Ns/m².

Solution

Flow rate $Q = Av$
$= \pi d^2 v/4$

Thus $v = 4Q/\pi d^2$
$= 4 \times 0.1/\pi \times 0.3^2$
$= 1.41$ m/s

Head loss $h_f = \dfrac{\lambda l}{d} \dfrac{v^2}{2g}$

$= \dfrac{0.018 \times 400}{0.3} \times \dfrac{1.41^2}{2g}$

$= 2.43$ m

Applying the Bernoulli equation

$$z_1 + \frac{p_1}{\rho g} + \frac{v_1^2}{2g} = z_2 + \frac{p_2}{\rho g} + \frac{v_2^2}{2g} + \text{head loss}$$

to point 1 at the inlet and point 2 at the outlet gives

$5 + \dfrac{200 \times 10^3}{10^3 g} = \dfrac{p_2}{10^3 g} + 2.43$

since $v_1 = v_2$

Thus $5 + 20.39 = \dfrac{p_2}{10^3 g} + 2.43$

or $p_2 = 22.96 \, g \times 10^3$
$= 225 \times 10^3$ N/m²
$= 225$ kN/m²

Reynolds number $Re = \dfrac{\rho d v}{\mu}$

$$= \frac{10^3 \times 0.3 \times 1.41}{1.14 \times 10^{-3}}$$

$$= 3.71 \times 10^5$$

This shows that the flow is turbulent since $Re > 4000$.

6.9 Flow in open channels – the Chezy formula

This formula gives the mean flow velocity in an open channel with constant cross-sectional area when the flow is incompressible, steady, uniform and turbulent.

Consider a section ABCD of uniform flow (that is, at constant depth) in a sloping channel with constant cross-sectional area A as in Fig. 6.7.

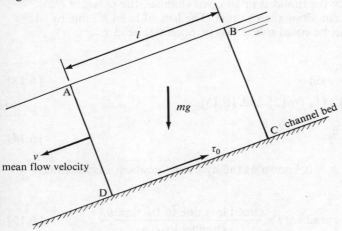

Fig. 6.7 Notation for Chezy formula

Let the length of the section be l and the channel bed inclination to the horizontal be θ. If the length of the perimeter of the channel in contact with the liquid is P and the shear stress is τ_0 at the channel surface then the shear force on the section ABCD is

shear force = shear stress × wetted area of contact
$$= \tau_0 \times P \times l$$

Since there is no acceleration, the shear force equals the component of the weight of the section ABCD in the direction parallel to the bed slope.

This is $mg \sin \theta$ or $\rho A l g \sin \theta$
Thus $\tau_0 P l = \rho A l g \sin \theta$

As before, assume $\tau_0 = kv^2$ [6.8]
Thus $kv^2 P = \rho A g \sin \theta$

or $v^2 = \dfrac{\rho g}{k} \dfrac{A}{P} \sin \theta$ [6.11]

The ratio A/P is known as the hydraulic mean depth m (sometimes denoted by R). Hence

hydraulic mean depth $m = \dfrac{\text{area of cross-section of flow } A}{\text{wetted perimeter } P}$ [6.12]

As the liquid flows down the channel, the mean flow velocity is constant. Since the liquid is in an open channel, the pressure also remains constant along the channel. The loss of head h_f, due to friction, must thus be equal to the loss of potential head z.

Thus $z = h_f$
But $z = l \sin \theta$
and hence $h_f = l \sin \theta$ [6.13]

Thus from [6.11], [6.12] and [6.13],

$v^2 = \dfrac{\rho g}{k} m \dfrac{h_f}{l}$ [6.14]

The ratio h_f/l is known as the hydraulic gradient i (sometimes denoted by S).

Thus hydraulic gradient $i = \dfrac{\text{head loss due to friction } h_f}{\text{channel length } l}$ [6.15]

From [6.13], h_f/l is just $\sin \theta$, that is, i is the slope of the channel bed.

Finally, using [6.14] and [6.15],

$v^2 = \dfrac{\rho g}{k} mi$

or $v = \sqrt{\dfrac{\rho g}{k}} \sqrt{mi}$

or $v = C\sqrt{mi}$ [6.16]

This is the Chezy formula and $C = \sqrt{\dfrac{\rho g}{k}}$ is called the Chezy coefficient. For turbulent flow C depends on the area and shape of the channel section and on the roughness of its surface. Its units are $m^{1/2}/s$.

6.10 Reynolds number for channel flow

For pipe flow the Reynolds number Re is defined as the ratio $\rho dv/\mu$ – [6.6]. For flow in a channel, the diameter d is replaced by m, the hydraulic mean depth as defined by [6.12]. For channel flow then

$$Re = \frac{\rho m v}{\mu} \qquad [6.17]$$

If Re is less than 500 then the flow in a channel is likely to be laminar. If it is greater than about 1000 flow is likely to be turbulent.

Worked example 6.7
A canal has a rectangular cross-section. The width is 3 m and the water depth is 1.2 m. If the bed slope is 1 in 1000 and the Chezy coefficient is 50 $m^{1/2}/s$, estimate the flow rate in the channel.

Solution

Hydraulic mean depth $m = \dfrac{A}{p} = \dfrac{3 \times 1.2}{3 + 2 \times 1.2} = 0.667$ m

$i = 1/1000 = 0.001$

Mean flow velocity $v = C\sqrt{mi}$
$ = 50 (0.667 \times 0.001)^{\frac{1}{2}}$
$ = 1.29$ m/s

Flow rate $Q = Av$
$ = 3 \times 1.2 \times 1.29$
$ = 4.64$ m³/s

Worked example 6.8
A culvert has trapezoidal section with a base width of 2.5 m and sides sloping outwards at 45° to the horizontal. If the water depth in the culvert is 1.5 m and the Chezy coefficient is 82 SI units, estimate the flow rate in m³/day, when the channel bed slope is 1 in 3000.

Solution
Top width at water level = 2.5 + 2(1.5) = 5.5 m

Hydraulic mean depth $m = \dfrac{A}{P} = \dfrac{\frac{1}{2}(5.5 + 2.5)1.5}{2.5 + 2 \times 1.5 \times \sqrt{2}}$

$ = 0.890$ m

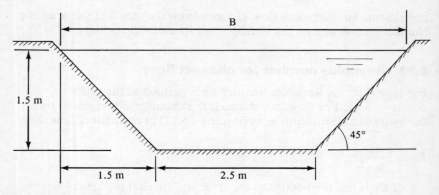

Fig. 6.8 Worked example 6.8

Mean flow velocity $v = C\sqrt{mi}$
$= 82\,(0.890/3000)^{\frac{1}{2}}$
$= 1.41$ m/s

Flow rate $Q = Av$
$= \frac{1}{2}(5.5 + 2.5)1.5 \times 1.41$ m³/s
$= \frac{1}{2} \times 8 \times 1.5 \times 1.41 \times 3600 \times 24$ m³/day
$= 7.31 \times 10^5$ m³/day

Exercises

Objective type
Choose the ONE response which is the most appropriate.

6.1 For incompressible flow the continuity equation may be written as
  A $Q = \rho v$
  B $A_1 v_1 = A_2 v_2$
  C $Q = \rho v A$
  D $p_1 A_1 = p_2 A_2$

6.2 If a pipe has cross-sectional area 1000 mm² and the flow velocity is 0.2 m/s the flow rate is
  A 2×10^{-4} m³/s
  B 2 litre/s
  C 2×10^{-2} m³/s
  D 20 litre/s

6.3 In the equation $z + \dfrac{p}{\rho g} + \dfrac{v^2}{2g} = $ constant the units for each term could be

    A kN/m²
    B m²/s
    C kN/m³s
    D m
6.4 The pressure head is
    A $g\rho z$
    B $v^2/2g$
    C $p/\rho g$
    D p
6.5 In laminar flow
    A the fluid must have a high density
    B the fluid particles move in an orderly manner
    C the fluid must have no viscosity
    D the fluid must have a low viscosity
6.6 An ideal fluid
    A is very viscous
    B is incompressible
    C has no surface tension
    D has no viscosity
6.7 In turbulent flow
    A the fluid moves in layers sliding over each other
    B the fluid particles move along irregular paths
    C the fluid must move with a high velocity
    D the fluid must have a high viscosity
6.8 Turbulent flow in a pipe is most likely when
    A the fluid is very viscous
    B the flow is in a large diameter pipe
    C the flow velocity is high
    D a combination of B and C
6.9 Laminar flow in a pipe is most likely when
    A the flow is in a small diameter pipe
    B the flow velocity is low
    C a combination of A and B
    D the fluid viscosity is low
6.10 Steady flow occurs
    A only for ideal fluids
    B when the pressure and velocity at a point in the flow do not change with time
    C when the pressure and velocity at a point in the flow change slowly with time
    D when the velocity at all points in the flow is the same
6.11 Uniform flow occurs
    A when the velocity at all points in the flow is the same
    B when the flow is laminar
    C when flow takes place through a tapering pipe
    D when the flow is steady

6.12 The units for the friction coefficient λ in Darcy's formula is
  A m
  B m/s
  C N/m^2
  D none of these

6.13 Hydraulic mean depth could be measured in
  A m^3
  B m^2
  C m
  D none of these

6.14 For turbulent flow in a channel, the value of the Chezy coefficient depends on
  A the area of cross-section
  B the shape
  C the pipe roughness
  D all of these

6.15 For a circular pipe with diameter d running half-full, the hydraulic mean depth is
  A $d/4$
  B $d/3$
  C $d/2$
  D d

Problems

6.16 The mean flow velocity in a 100 mm diameter water pipe is 1.2 m/s. Find the flow rate in m^3/s, litre/s and kg/s.

6.17 Oil leaves a tank at a rate of 200 litres every 10 s, through a 200 mm diameter pipe. What is the mean flow velocity in the pipe?

6.18 A water pipe AB has diameter 50 mm. The flow rate is 8.7 litre/s. The pipe AB is in series with a pipe BC of diameter 75 mm. At C the pipe branches into two pipes, CD with diameter 70 mm and CE with diameter 30 mm. Find the mean flow velocities in all the pipes if the flow rate in CD is twice that in CE.

6.19 Water flows down a vertical pipe section 3 m long. The flow rate is 0.05 m^3/s. The top and bottom diameters of the section are 150 mm and 75 mm respectively. Find the pressure difference between the top and bottom of the pipe section if all losses are negligible.

6.20 A water pipeline consists of a section AB, of diameter 100 mm in series with a section BC of diameter 75 mm. The pressure and mean flow velocity at A are 300 kN/m^2 and 2 m/s respectively. The elevation of A is 4 m above that of C. Find the mean flow velocity and pressure at C, neglecting all losses.

6.21 A steel pipe of length 400 m and diameter 300 mm connects two reservoirs. If the friction factor is 0.016, find the head loss due to friction when the flow rate is 0.06 m^3/s.

6.22 The head available at the entrance of a pipeline 800 m long is 20 m. The friction coefficient is 0.012. Find the minimum pipe diameter required if the discharge rate is to be 0.5 m³/s. Ignore losses other than friction.

6.23 A drainage pipe, when flowing full, has to discharge 5000 litre/min of water. The head loss must not exceed 5 m/km of pipeline. If the friction factor is 0.017, find the pipe diameter required.

6.24 A rectangular channel is 2.7 m wide. The bed slope is 0.001 and the flow depth is 1.8 m. Find
(a) the hydraulic mean depth;
(b) the flow velocity;
(c) the volume flow rate.
The Chezy coefficient is 70 SI units.

6.25 An open channel has a semi-circular cross-section. The radius is 1.0 m and the channel runs full. If the bed slope is 0.04%, find the volume rate of flow. The Chezy coefficient is 55 SI units.

6.26 A drainage channel has a horizontal bed of width 3 m and side slopes of 1 vertical to 2 horizontal. The bed slope is 2×10^{-4}. If the water depth is 1.2 m and the Chezy coefficient is 40 SI units, find the discharge rate.

6.27 A rectangular canal is to be designed to carry 10 m³/s of water at an average flow velocity of 0.8 m/s when running full. If the width is to be twice the depth, find suitable dimensions and bed slope for the canal assuming the Chezy coefficient to be 30 SI units.

Chapter 7

Fluid measurements

Learning objectives

After reading this chapter and working through the exercises you should be able to:
- measure pressure and pressure difference using a piezometer and a manometer;
- measure fluid flow by the use of Venturi meters, notches and weirs;
- measure energy losses in straight pipes, bends and elbows.

7.1 Pressure measurement – the piezometer

The simplest way to measure pressure in a pipe is to use a piezometer. This is simply a length of vertical, transparent tubing. This method of pressure measurement was first used by Bernoulli. The piezometer may be connected into a pipeline as shown in Fig. 7.1.

The height to which the liquid rises is a measure of the pressure. The gauge pressure at B is zero since B is at atmospheric pressure. The gauge pressure p at the pipe axis A is thus given by

$$p = g\rho h \qquad [7.1]$$

If the liquid in the pipe is flowing, care must be taken to ensure that there is no roughness or projection into the flow round the opening where the piezometer is connected into the pipe. Ideally, the opening from the pipe should have a slightly rounded edge. If the

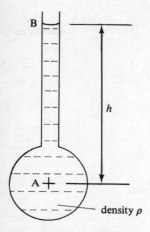

Fig. 7.1 A piezometer

piezometer has a very small diameter, surface tension effects will cause an error in the value of h. If the tube diameter is greater than about 10–15 mm then the error will probably be negligible.

The piezometer is simple and inexpensive. However, it has several disadvantages; for example, it cannot be used to measure gas pressure or for negative liquid gauge pressures. Also the length of vertical tube which can conveniently be used limits the maximum pressure which may be measured.

Worked example 7.1

The water level in a piezometer on a water pipe is 0.5 m above the pipe axis. Find the gauge pressure at the pipe axis in kN/m².

Solution

Gauge pressure p at pipe axis is given by

$p = g\rho h$ (see Fig. 7.1)
$ = 9.81 \times 10^3 \times 0.5$
$ = 4.90 \times 10^3$ N/m²
$ = 4.90$ kN/m²

7.2 The simple U–tube manometer

This is shown in Fig. 7.2 measuring the pressure in a pipe. The liquid in the manometer may or may not be the same as the fluid in the pipe. Manometers like this have been used since 1662 when one was used by Robert Boyle, remembered for his famous gas law.

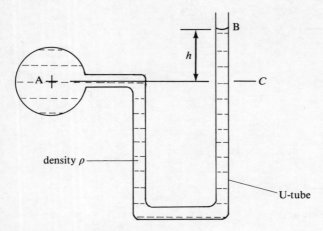

Fig. 7.2 A U-tube manometer

The U-tube and pipe are both full of the same liquid. Gauge pressure at B is zero, and hence gauge pressure at C = $g\rho h$. Points A and C are on the same horizontal line and thus $p_A = p_C$. Thus

gauge pressure p_A at the pipe axis = $g\rho h$. [7.2]

The absolute pressure at A would be $g\rho h + p_{ATM}$ where p_{ATM} is atmospheric pressure.

If the arrangement in Fig. 7.2 is to be used to measure gas pressures then a liquid must be used in the manometer tube. This liquid must not mix or react chemically with the gas in the pipe — mercury is widely used for large pressures and oil for small pressures. A U-tube manometer is shown in Fig. 7.3 measuring the gas pressure in a pipe — this arrangement can also be used when the pipe fluid is a liquid, provided, of course, that pipe and manometer liquids do not mix or react chemically.

The manometric liquid density ρ_m must be greater than the pipe fluid density ρ.

Gauge pressure at E $p_E = 0$
Gauge pressure at D $p_D = g\rho_m h$

If gauge pressures at A and B are p_A and p_B respectively, then $p_A = p_B$ since A and B are points on the same horizontal level. Further,

Gauge pressure at C $p_C = p_A + g\rho a$

Since C and D are on the same horizontal level $p_C = p_D$. Thus

$p_A + g\rho a = g\rho_m h$
or $p_A = g(\rho_m h - \rho a)$ [7.3]

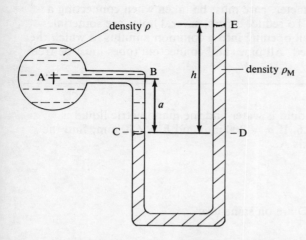

Fig. 7.3 U-tube manometer – positive gauge pressure

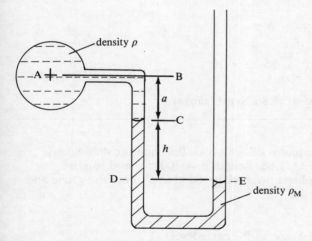

Fig. 7.4 U-tube manometer – negative gauge pressure

In the case of a gas pipe ρ will be very much less than ρ_m and the ρa term may then usually be neglected.

In Fig. 7.3, level E is above level C. If E is below C then the situation would be as illustrated in Fig. 7.4.

It is left as a simple exercise for the reader to show that in this case

$$p_A = -g(\rho_m h + \rho a) \qquad [7.4]$$

In this case, the pressure at A is negative gauge.

As with the piezometer, care must be taken when connecting a manometer to a pipe. To reduce errors, several holes are sometimes used round the pipe, all opening into a common annulus to which the manometer is connected. All pipes and connecting tubes must be free of air or gas bubbles.

Worked example 7.2
In Fig. 7.3, the pipe liquid is water and the manometric liquid is mercury with S.G. 13.6. If $a = 200$ mm and $h = 650$ mm, find the pressure at the pipe axis A.

Solution
Refer to Fig. 7.3.

$p_C = p_D$ since C and D are on same level

$p_D = g\rho h$
$= 9.81 \times 10^3 \times 13.6 \times 0.65$
$= 86.72 \times 10^3$

$p_C = p_A + g\rho a$
$= p_A + 9.81 \times 10^3 \times 0.2$
$= p_A + 1.962 \times 10^3$

Thus $p_A + 1.96 \times 10^3 = 86.72 \times 10^3$
and $p_A = 84.8 \times 10^3$ N/m^2

that is, pressure at A $= 84.8$ kN/m^2 (gauge)

Worked example 7.3
In Fig. 7.5 the pipe liquid is oil with S.G. 0.83 and the manometric liquid is brine with S.G. 1.08. Length $a = 100$ mm and length $h = 55$ mm. Atmospheric pressure is 100 kN/m^2. Find the gauge and absolute pressures at A.

Solution
$p_X = p_Y$ since X and Y are on the same level

Now $p_Y = 0$ (gauge)
and $p_Z = p_A + g\rho_o a$

and hence $p_X = p_Z + g\rho_b h = p_A + g\rho_o a + g\rho_b h$

that is $p_X = p_A + 9.81 \times 0.83 \times 10^3 \times 0.1 + 9.81 \times 1.08 \times 10^3 \times 0.055$
$= p_A + 1.40 \times 10^3$ (gauge)

Thus $0 = p_A + 1.40 \times 10^3$

or $p_A = -1.40 \times 10^3$ N/m^2
$= -1.40$ kN/m^2 (gauge)

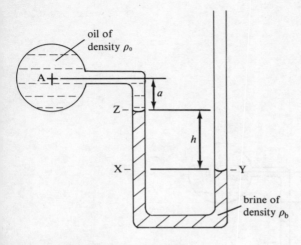

Fig. 7.5 Worked example 7.3

But $p_{ABS} = p_{gauge} + p_{ATM}$ where p_{ATM} is atmospheric pressure

Thus $p_A = -1.40 + 100$
$= 98.6$ kN/m² (absolute)

The gauge and absolute pressures at A are thus -1.43 kN/m² and 98.6 kN/m² respectively.

7.3 The differential manometer

If the pressure difference between two points is required, then the two limbs of the simple U-tube manometer may be connected to the two points. This is illustrated in Fig. 7.6 where the pressure difference between two pipes is being measured. The manometer used in this way is known as a differential manometer. All connecting pipes must be full of pipe fluid.

The use of a differential manometer can best be understood by considering a numerical example.

Worked example 7.4
In Fig. 7.6, the two pipes both contain water and the manometer liquid is mercury with S.G. 13.6. The axis of pipe A is 0.75 m vertically above the axis of pipe B. The difference in the levels in the manometer is 0.6 m. Find the pressure difference between the pipes.

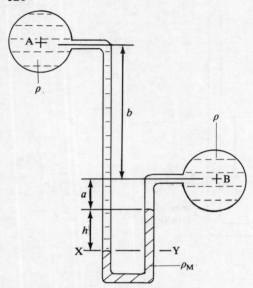

Fig. 7.6 Differential manometer

Solution
Refer to Fig. 7.6.
Two points on the same horizontal in the same mass of fluid must be chosen. These can be X and Y. Since pressures at X and Y are equal,

$p_X = p_Y$

Now $p_X = p_A + g\rho (b + a + h)$
$= p_A + 9.81 \times 10^3 (0.75 + a + 0.6)$
$= p_A + 13.24 \times 10^3 + 9810 a$

and $p_Y = p_B + g\rho a + g\rho_m h$
$= p_B + 9810 a + 9.81 \times 13.6 \times 10^3 \times 0.6$
$= p_B + 9810 a + 80.05 \times 10^3$

Thus $p_A + 13.24 \times 10^3 + 9810 a = p_B + 9810 a + 80.05 \times 10^3$
or $p_A - p_B = (80.05 - 13.24) \times 10^3$ N/m²

(note that the length a is not required)

Thus $(p_A - p_B)$, the pressure difference between pipes A and B, is 66.8×10^3 N/m² or 66.8 kN/m²

7.4 Inverted differential manometer

If the differential manometer is used to measure small pressure differences, then the manometric liquid may be a light oil or air. If the manometric fluid has a density less than the density of the fluid under test, then the U-tube must be inverted as shown in Fig. 7.7 to prevent the fluid in it being swept away. The tap T can be used to adjust the levels in the manometer or to remove air bubbles.

If the pressure drop across a measured length of pipe is found, then, if the flow velocity is known, a value can be found for the friction factor λ as illustrated in the following example.

Worked example 7.5

An inverted U-tube manometer is used to measure the pressure drop between two points A and B 2 m apart in a horizontal 100 mm diameter water pipe as shown in Fig. 7.7. The liquid in the manometer is oil with S.G. 0.88 and the difference in levels in the manometer is 20 mm. Find the pressure difference between A and B. If the flow velocity in the pipe is 0.4 m/s, what is the friction factor for the pipe?

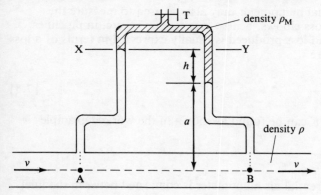

Fig. 7.7 Inverted differential manometer

Solution

Refer to Fig. 7.7. Since X and Y are on the same level,

$p_X = p_Y$

Now $p_X = p_A - g\rho (h + a)$ (note the negative sign)
        $= p_A - 9.81 \times 1000 (0.02 + a)$

and $p_Y = p_B - g\rho a - g\rho_m h$
        $= p_B - 9.81 \times 1000 \times a - 9.81 \times 0.88 \times 1000 \times 0.02$

Thus $p_A - 9.81 \times 1000 (0.02 + a) = p_B - 9.81 \times 1000 \times a - 9.81 \times 0.88 \times 1000 \times 0.02$

and $p_A - p_B = 196.2 - 172.7$ (note that the value of a is not required)
$= 23.5$ N/m²

Hence the pressure difference between A and B is 23.5 N/m²

The head loss $h_f = (p_A - p_B)/\rho g = 23.5/10^3 g = 2.40 \times 10^{-3}$ m. However, by the Darcy–Weisbach formula, [6.10],

$$h_f = \frac{\lambda l}{d} \frac{v^2}{2g}$$

Thus $2.40 \times 10^{-3} = \dfrac{\lambda \times 2}{0.1} \times \dfrac{0.4^2}{2g}$

or $\lambda = 2.4 \times 10^{-3} \times 0.1 \times 2g/2 \times 0.4^2 = 0.0147$

The friction factor is thus 0.0147.

The differential manometer may also be used to measure the pressure drop across elbows, bends, meters, valves, etc, in pipelines. For these, the head loss produced is usually expressed in terms of a loss coefficient K where

$$\text{head loss} = K\frac{v^2}{2g} \qquad [7.5]$$

The value of K can be found as shown in the worked example below.

Worked example 7.6
An inverted differential air manometer is connected between the inlet and outlet of a 90° bend on a 75 mm diameter horizontal water pipeline. The difference in levels is 30 mm. A meter in the pipe records the flow as 4 litre/s. Find the loss coefficient K for the bend.

Solution
Since X and Y are on the same level

$p_X = p_Y$

Thus $p_{INLET} - (a + h)\rho g = p_{OUTLET} - \rho g a - \rho_{air} g h$

Since ρ_{air} is negligible compared with ρ

$p_{INLET} - p_{OUTLET} = g\rho h$ since the $\rho g a$ terms cancel

Thus head loss $= (p_{INLET} - p_{OUTLET})/g\rho = h = 30$ mm

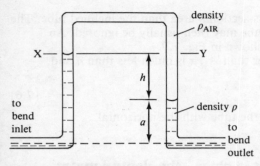

Fig. 7.8 Worked example 7.6

But head loss = $K \dfrac{v^2}{2g}$

and $v = Q/A$

$$= \dfrac{4 \times 10^{-3} \times 4}{\pi \times 0.075^2}$$

$$= 0.905 \text{ m/s}$$

Thus $K = \dfrac{30 \times 10^{-3} \times 19.62}{0.905^2}$

$$= 0.719$$

The loss coefficient of the bend is thus 0.719.

7.5 The inclined tube manometer

For small pressure differences, the accuracy of reading may be increased by inclining one limb of the U-tube. To remove the necessity to read the meniscus level on both sides of the manometer, the other limb may

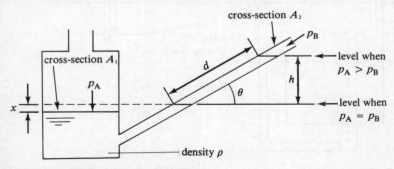

Fig. 7.9 Inclined-tube manometer

be made much larger in cross-sectional area than the inclined tube. The change in level in the wide tube may then usually be ignored. An inclined tube manometer is shown in Fig. 7.9.

When A_1 is much greater than A_2, x is much less than d and, approximately,

$$p_A - p_B = g\rho h = g\rho d \sin\theta \qquad [7.6]$$

where θ is the inclination of the tube with the horizontal.

7.6 Flow measurement in pipes – the Venturi meter

The simplest direct method for finding the volume flow rate in a pipe is to weigh the liquid flowing in a measured time. The volume flow rate can then be found using the liquid density.

The best known indirect method for finding flow rate is probably using the Venturi meter. This consists of a converging inlet section followed by a short cylindrical section known as the throat and a diverging outlet section. A Venturi meter is shown in Fig. 7.10 in a horizontal pipe. The length and diameter of the throat are often equal. Typical values for the cones of convergence and divergence are 7° and 20° respectively. The divergence is more gradual than the convergence to reduce the head loss across the meter.

As the liquid flows through the throat, its velocity is greater than in the pipe. By the Bernoulli theorem, the pressure in the throat must therefore be less than in the pipe. The pressure difference between throat and pipe can be used to deduce the flow rate.

If p_0 and p_c are the pressures in the pipe and throat respectively, then using the Bernoulli theorem applied to points on the pipe axis and

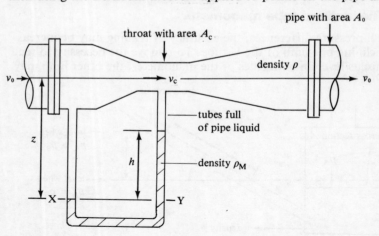

Fig. 7.10 Venturi meter

throat axis,

$$\frac{p_0}{\rho g} + \frac{v_0^2}{2g} = \frac{p_c}{\rho g} + \frac{v_c^2}{2g}$$

since $z_0 = z_c$ for a horizontal pipe.

By the continuity equation

$$Q = A_0 v_0 = A_c v_c$$

Hence $v_0 = \dfrac{A_c}{A_0} v_c$

Thus $\dfrac{p_0}{\rho} + \left(\dfrac{A_c}{A_0}\right)^2 \dfrac{v_c^2}{2} = \dfrac{p_c}{\rho} + \dfrac{v_c^2}{2}$

or $\dfrac{v_c^2}{2}\left[1 - \left(\dfrac{A_c}{A_0}\right)\right]^2 = \dfrac{p_0 - p_c}{\rho}$

or $v_c = \left[\dfrac{2(p_0 - p_c)/\rho}{1 - (A_c/A_0)^2}\right]^{\frac{1}{2}}$

and thus $Q = A_c v_c = A_c \left[\dfrac{2(p_0 - p_c)/\rho}{1 - (A_c/A_0)^2}\right]^{\frac{1}{2}}$ [7.7]

This is the ideal flow rate. The actual flow rate is slightly less than this due to head loss introduced by the meter. To allow for this, [7.7] is rewritten as

$$Q = C_D A_c \left[\frac{2(p_0 - p_c)/\rho}{1 - (A_c/A_0)^2}\right]^{\frac{1}{2}} \qquad [7.8]$$

where C_D is the discharge coefficient. It has value typically between 0.94 and 0.98. The value of C_D varies slightly with the ratio A_c/A_0 and with the Reynolds number. Its value may be found by experiment.

If the pressure difference is found using a differential manometer as shown in Fig. 7.10, then at the points X and Y

$p_X = p_Y$
or $p_0 + g\rho z = p_c + g\rho(z - h) + g\rho_m h$

that is $p_0 - p_c = gh(\rho_m - \rho)$ [7.9]

Hence using [7.8] and [7.9]

$$Q = C_D A_c \left[\frac{2gh(\rho_m/\rho - 1)}{1 - (A_c/A_0)^2}\right]^{\frac{1}{2}} \qquad [7.10]$$

The value of Q can thus be found from a measurement of h. Equation

[7.10] applies even when the Venturi meter is inclined at an angle to the horizontal.

In practice the throat diameter should not be made too small otherwise the pressure in the throat falls so low that dissolved gases are released from the liquid. These can lead to inaccurate readings especially if bubbles enter the manometer tubes. For water, the minimum throat pressure is about 2 m of water absolute.

To eliminate errors due to flow disturbances caused by bends, etc, there should be a straight length of pipe upstream of the Venturi meter of length at least twenty pipe diameters.

Worked example 7.7
A Venturi meter in a horizontal oil pipeline with diameter 250 mm has a throat diameter of 160 mm. The pressure difference between pipe and throat is measured using a mercury differential manometer and the level difference in this is 162 mm. If the discharge coefficient for the meter is 0.97 and the S.G. of the oil is 0.9, find the flow rate in the pipe in litre/s.

Solution
From [7.10]

$$Q = C_D A_c \left[\frac{2gh\,(\rho_m/\rho - 1)}{1 - (A_c/A_0)^2} \right]^{\frac{1}{2}}$$

$$A_c = \pi d_c^2/4$$
$$= \pi \times 0.160^2/4$$
$$= 2.02 \times 10^{-2} \text{ m}^2$$

$$(A_c/A_0)^2 = (d_c/d_0)^4$$
$$= (0.160/0.250)^4$$
$$= 0.168$$

Thus $Q = 0.97 \times 2.02 \times 10^{-2} \left[\dfrac{2 \times g \times 0.162\,(13.6/0.9 - 1)}{1 - 0.168} \right]^{\frac{1}{2}}$

$$= 0.144 \text{ m}^3/\text{s}$$
$$= 144 \text{ litre/s}$$

The flow rate is thus 144 litre/s.

7.7 Flow measurement in channels – notches and weirs

The flow rate in a channel may be determined using a weir by finding the depth or head of water above the top of the weir. There are different types of weir. These include the sharp-crested (or thin plate) weir and the broad-crested weir. The sharp-crested weir consists of a thin vertical metal (brass or steel) plate with a bevel on the downstream side. The two commonest shapes in use are rectangular and triangular (also called V-notch). The crest or sill should be shaped as in Fig. 7.11.

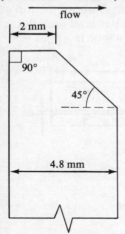

Fig. 7.11 Standard sharp-crested (thin-plate) weir

Disadvantages of this type of weir are that it is easily damaged by floating debris, the sharp edge becomes rounded with continuous use and it tends to silt up. Its use is limited to small streams and to laboratory measurements.

The broad-crested weir has an appreciable length in the direction of flow. It may have a rectangular section and normally spans the full width of a channel.

7.8 The rectangular sharp-crested weir

Flow over the crest of a sharp-crested rectangular weir is shown in Fig. 7.12.

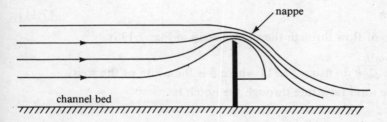

Fig. 7.12 Flow over a weir

The level falls as the water approaches the weir and the streamlines curve upwards. The overflowing sheet of water is known as the nappe. For accurate flow measurement the air pressure under the nappe must be atmospheric. To ensure this, air vents may have to be provided.

To obtain a formula for Q in terms of the head H above the top of the weir as defined in Fig. 7.13, assume that the water level does not fall near the weir and that the pressure within the nappe is atmospheric.

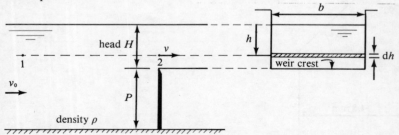

Fig. 7.13 Notation for flow over a rectangular weir

For an ideal fluid, the total head must be a constant by the Bernoulli theorem. Application of the theorem to points 1 and 2 in Fig. 7.13 gives

$$\frac{p_1}{\rho g} + \frac{v_0^2}{2g} = \frac{p_2}{\rho g} + \frac{v^2}{2g}$$

since the points are on the same level.

However $p_1 = g\rho h$
and $p_2 = 0$ since the pressure in the nappe is atmospheric.

Thus $h + \dfrac{v_0^2}{2g} = \dfrac{v^2}{2g}$

If, as is often the case in practice, the upstream (or approach) velocity v_0 is negligible compared with v, then

$$v = \sqrt{2gh} \qquad [7.11]$$

The rate of flow through the shaded strip in Fig. 7.13 is

$dQ = vb dh$
$ = b\sqrt{2gh}\, dh$ using [7.11] where b is the width of the weir.

Thus the total flow rate through the notch is

$$Q = \int dQ = b\sqrt{2g} \int_0^H h^{\frac{1}{2}}\, dh$$

or $Q = \frac{2}{3} b \sqrt{2g} \left[h^{3/2} \right]_0^H$

or $Q = \frac{2}{3} b \sqrt{2g} H^{3/2}$ (ideally) [7.12]

The actual flow rate is considerably less than this and to allow for this a discharge coefficient C_D is included in [7.12] giving

$$Q = \frac{2}{3} b C_D \sqrt{2g} H^{3/2}$$ [7.13]

The value of C_D, which varies (mainly with H and P) is about 0.6.

In practice, H should be measured a distance of at least four times the maximum head upstream from the weir.

Worked example 7.8

A rectangular weir in a horizontal channel is 2.5 m wide and the sill is 0.75 m above the channel bed. If C_D is 0.62 and the flow rate is 0.9 m³/s find the depth upstream from the weir.

Solution

$Q = \frac{2}{3} b C_D \sqrt{2g} H^{3/2}$ by [7.13]

Hence $H^{3/2} = \dfrac{3 \times 0.9}{2 \times 2.5 \times 0.62 \times \sqrt{2g}}$

$= 0.1966$

Hence $H = 0.338$ m

The upstream depth is thus $(0.338 + 0.75)$ m, that is, 1.09 m.

7.9 The triangular sharp-crested weir

By a method similar to that in Section 7.8, it may be shown that the flow rate is

$$Q = \frac{8}{15} C_D \sqrt{2g} \tan\frac{\theta}{2} H^{5/2}$$ [7.14]

where θ is the angle of the weir as shown in Fig. 7.14 and C_D, the discharge coefficient is again about 0.6 although its value depends on θ and H.

Since $Q \propto H^{5/2}$ for the triangular (V-notch) weir, it is preferable to the rectangular notch for small values of Q because it is more sensitive to changes in H.

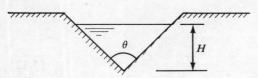

Fig. 7.14 Notation for flow over a triangular weir

Worked example 7.9
Find the flow rate for a 90° triangular weir when the head is 0.75 m. The discharge coefficient is 0.585.

Solution

$$Q = \frac{8}{15} C_D \sqrt{2g} \tan \frac{\theta}{2} H^{5/2} \quad \text{by [7.14]}$$

$$= \frac{8}{15} \times 0.585 \sqrt{2g} \tan 45° (0.75)^{5/2}$$

$$= 0.673 \text{ m}^3/\text{s}$$

The flow rate is thus 0.673 m³/s.

7.10 The broad-crested weir

This type of weir has an appreciable length in the direction of flow. Fig. 7.15 shows the flow over a rectangular profile broad-crested weir with rounded edges.

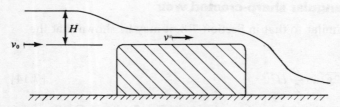

Fig. 7.15 Flow over a broad-crested weir

When the depth downstream is less than the crest height, flow over the weir is likely to be unaffected by the conditions downstream.

It may then be shown that when the approach velocity v_0 is small compared with v, the flow rate Q over the weir is given approximately by

$$Q = \left(\frac{2}{3}\right)^{3/2} C_D B \sqrt{g}\, H^{3/2}$$

or $Q = 1.70\, C_D B H^{3/2}$ [7.15]

where B is the width of the weir and C_D is the discharge coefficient whose value depends on the weir geometry and on H.

Worked example 7.10
Water flows over a broad-crested weir in a rectangular channel 2 m wide. The crest of the weir is 500 mm above the channel bed and the upstream depth is 700 mm. Estimate values for the channel flow rate and the approach velocity if the discharge coefficient is 0.85.

Solution
Assuming that the approach velocity is negligible

$Q = 1.70\, C_D B H^{3/2}$ by [7.15]

However $H = (700 - 500) = 200$ mm

Thus $Q = 1.70 \times 0.85 \times 2 \times 0.2^{3/2}$
   $= 0.26$ m^3/s

Approach velocity is given by

$Q = Av$

where $A = 0.7 \times 2 = 1.4$ m^2

Thus $v = \dfrac{0.26}{1.4}$

   $= 0.19$ m/s

Exercises

Objective type
Choose the ONE response which is the most appropriate.

7.1 A piezometer tube can conveniently be used to measure
   A small pressures in gas pipes
   B small positive gauge pressures in water pipes
   C small negative gauge pressures in water pipes
   D large positive gauge pressures in water pipes

7.2 The water level in a piezometer on a pipe is 30 mm above the pipe axis. The gauge pressure at the pipe axis is
   A -30 mm water
   B 294 N/m^2
   C 294 kN/m^2
   D none of these

7.3 The pressure at A in the water pipe shown in Fig. 7.16, in kN/m^2, is
   A -1.96 gauge
   B 1.96 gauge
   C 1.96 absolute
   D -1.96 absolute

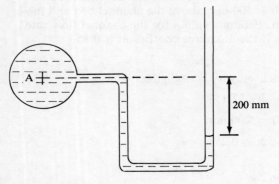

Fig. 7.16 Exercise 7.3

7.4 The pressure in the water pipe shown in Fig. 7.17 is measured using a mercury (S.G. = 13.6) manometer. The pressure at the pipe axis is
   A 0.5 m mercury gauge
   B 5.9 m water gauge
   C 57.9 kN/m^2 absolute
   D 1 m water gauge

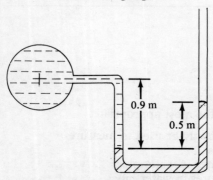

Fig. 7.17 Exercise 7.4

7.5 The sensitivity of an inclined-tube manometer can be increased by
   A   increasing the inclination with the horizontal
   B   using a liquid with a higher density
   C   decreasing the diameter of the inclined limb
   D   all of these

7.6 A typical value for the discharge coefficient C_D of a Venturi meter is
   A   0.5
   B   0.62
   C   0.98
   D   1.0

7.7 To prevent air bubbles being released from the water in a Venturi meter, the absolute pressure at the throat should not fall below about
   A   2 kN/m²
   B   200 kN/m²
   C   0.2 m of water
   D   2 m of water

7.8 To ensure uniform flow conditions, there should be a section of straight pipe upstream of a Venturi meter with length at least
   A   two pipe diameters
   B   2 m
   C   twenty pipe diameters
   D   20 m

7.9 The discharge rate over a V-notch weir is proportional to the head over the weir raised to the power
   A   1
   B   3/2
   C   2
   D   5/2

7.10 A typical value for the discharge coefficient of a rectangular weir is
   A   0.62
   B   0.98
   C   1.0
   D   none of these

7.11 The head loss across a bend in a water pipe is 25 mm of water and the mean flow velocity is 1 m/s. The loss coefficient of the bend is
   A   0.025
   B   0.49
   C   2.04
   D   none of these

Problems

7.12 What is the maximum gauge water pressure which could be measured using a piezometer tube 1.2 m high?

7.13 In Fig. 7.18, the manometer liquid is mercury with S.G. 13.6 and the pipe liquid is water. If h_1 is 0.30 m and h_2 is 0.43 m, what is the pressure at the pipe axis A?

7.14 In Fig. 7.18, the manometer liquid is water and the pipe fluid is a gas with density 1.24 kg/m³. Find the absolute pressure at A when atmospheric pressure is 100 kN/m², h_1 is 300 mm and h_2 is 200 mm. What would be the percentage error if the head of gas was neglected?

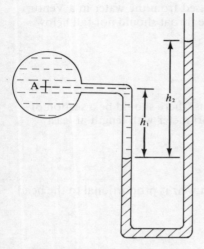

Fig. 7.18 Problems 7.13/7.14

7.15 In Fig. 7.19, the manometer liquid is mercury with S.G. 13.6 and the pipe liquids are both water. If h is 0.5 m, find the pressure difference between the pipes in kN/m² and in m of water.

7.16 An inverted U-tube containing air is used to measure the pressure difference between a tapping point A on a water pipe and another tapping point B 500 mm vertically below A as in Fig. 7.20. The water level difference in the U-tube is 800 mm. Find the pressure difference between the two points.

7.17 An inverted differential air manometer is used to measure the pressure difference between two points 1.8 m apart on a horizontal 75 mm diameter water pipe. The difference in water levels in the manometer is 40 mm. Find the pressure drop between the two points and hence find the friction coefficient for the pipe if the flow rate in the pipe is 4 litre/s. The density of air is negligibly small.

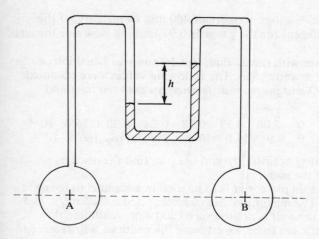

Fig. 7.19 Problem 7.15

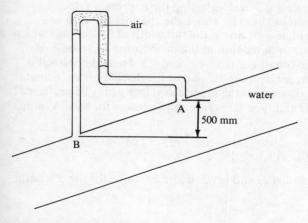

Fig. 7.20 Problem 7.16

7.18 A mercury manometer measures the pressure difference between the inlet and throat of a Venturi meter with throat diameter 50 mm in a horizontal 100 mm diameter water pipe. The difference in the mercury levels is 100 mm. Find the flow rate if C_D is 0.96. (S.G. of mercury is 13.6).

7.19 A Venturi meter with throat diameter 100 mm is located in a horizontal 300 mm diameter water pipeline. A pressure gauge at the inlet records 150 kN/m² gauge and a second gauge at the throat records a negative gauge pressure of 400 mm of mercury. If the discharge coefficient for the gauge is 0.97 find the flow rate through the meter.

7.20 A Venturi meter with throat diameter 16 mm was fitted into a 25 mm diameter water pipe. The following values were obtained for flow rate Q *and pressure difference* Δp between inlet and throat.

Q (m³/s)	0	0.06	0.12	0.18	0.24	0.30	0.36 × 10^{-3}
Δp (m water)	0	4.9	19.6	40.0	72.9	116	160 × 10^{-3}

Plot a graph of Q against $\sqrt{\Delta p}$ and use it to find the discharge coefficient for the meter.

7.21 A rectangular thin plate weir is to be used in a channel to measure the flow rate. The maximum discharge expected is 0.25 m³/s, and the head over the weir is not to exceed 200 mm. Assuming the discharge coefficient to be 0.6 estimate the width of weir required.

7.22 The head over the apex of a right-angled V-notch weir is 120 mm. If C_D is 0.62 what is the flow rate over the weir?

7.23 The mean flow velocity in a rectangular channel 2 m wide is 0.9 m/s and the flow depth is 0.3 m. Neglecting the approach velocity, estimate the height of the crest above the channel bed of a rectangular thin plate weir across the full width of the channel which would raise the upstream depth to 0.8 m. Assume C_D to be 0.63.

7.24 The height of the broad-crested weir across a 4 m wide channel is 0.5 m. The upstream depth is 1.2 m and the discharge coefficient is 0.9. What is the flow rate if the approach velocity may be neglected?

7.25 Values of head H and flow rate Q are shown below for a 90° V-notch weir.

H (mm)	25	50	75	100	125	150
Q (litre/s)	0.14	0.80	2.2	4.8	8.0	12.7

Plot log Q against log H and hence deduce a value for the discharge coefficient.

Part 4

Structural mechanics

Chapter 8

Pin-jointed frameworks

Learning objectives

After reading this chapter and working through the exercises you should be able to:
- list the assumptions made in the analysis of pin-jointed frameworks and explain the conditions required for a statically determinate framed structure;
- determine the nature and magnitude of forces in members of a pin-jointed framework (up to 17 members) subject to combined wind and dead loading by a graphical method;
- analyse frameworks, as above, by calculating the reactions, sketching the force diagram and applying trigonometric properties to establish the magnitude of the forces (that is, semi-graphical analysis).

Before considering the analysis of frameworks, some basic concepts will be revised.

8.1 Addition of forces – the triangle rule – resultant

Let two forces P and Q act at a point O as shown in Fig. 8.1(a).

To find the sum of the two forces, a line AB is first drawn to represent Q in magnitude and direction – Fig. 8.1(b). From B, a second line is then drawn to represent P in magnitude and direction. The line AC then represents the sum or resultant R of the two forces P and Q in magnitude and direction but not the position of R since R must act

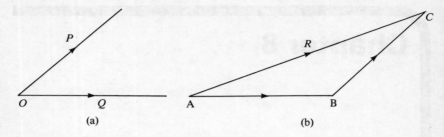

Fig. 8.1 Triangle rule

through O. This method of adding two forces is known as the triangle rule. The resultant of two or more forces may be described as that single force which produces the same effect as the individual forces and which can thus replace the individual forces. The magnitude and direction of R can be found by scale drawing or by calculation.

8.2 The parallelogram rule

The resultant of the forces P or Q in Fig. 8.1(a) can also be found using the parallelogram rule as illustrated in Fig. 8.2.

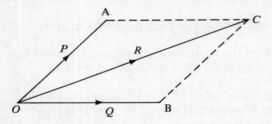

Fig. 8.2 Parallelogram rule

The method is to complete the parallelogram which has the lines representing P and Q as adjacent sides. The diagonal passing through the point O then represents the resultant R in magnitude and direction and in position.

8.3 Resolution of a force

Just as two forces may be added, so a single force may be split or resolved into two individual forces. These are known as the resolved parts or components of the force. This can be done in an infinite number of ways, but it is usual to resolve the force into two components at right-angles – see Fig. 8.3.

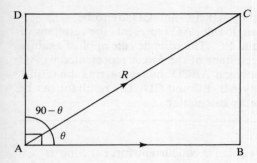

Fig. 8.3 Force components

Let a force R be represented by the line AC. The two components of R which are at right angles are then represented by AB and AD. The magnitudes of the components are $R \cos \theta$ for the component represented by AB and $R \sin \theta$ or $R \cos(90 - \theta)$ for the component represented by AD. This is equivalent to saying that the resolved part of any force R in a direction making an angle θ with the line of action of R is $R \cos \theta$.

8.4 Resultant of more than two coplanar forces – the polygon rule

Two forces may be added using the triangle rule. More than two forces may be added using the polygon rule. In this text only coplanar forces will be considered – that is, forces whose representative lines all lie in the same plane.

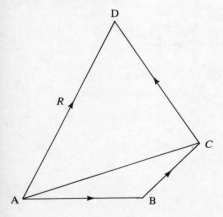

Fig. 8.4 Polygon rule

Suppose forces represented by AB, BC and CD are to be added—see Fig. 8.4. By the triangle rule, AC represents the resultant of the forces represented by AB and BC. The triangle rule applied again shows that AD represents the resultant of the forces represented by AC and CD. The side AD of the polygon ABCD thus represents the resultant of the forces represented by AB, BC and CD. The resultant can be determined by scale diagram or by calculation.

Worked example 8.1
Find the resultant of the three coplanar concurrent forces acting as shown in Fig. 8.5(a).

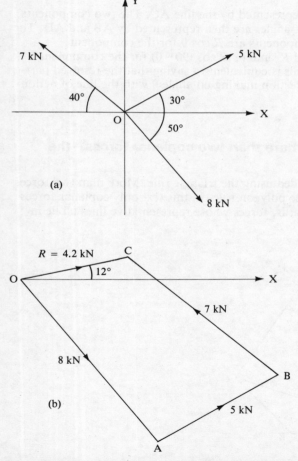

Fig. 8.5 Worked example 8.1 (a) Space diagram (b) Force diagram

Solution

Method 1: The polygon rule can be used. The three forces are represented by lines parallel to the three forces and with length proportional to their magnitudes. The resultant R is then represented by the line OC. The magnitude of R is 4.2 kN and its direction is 12° to OX see Fig. 8.5(b).

Method 2: The resultant can also be found by calculation. All the forces are resolved into their components along the directions OX and OY.

	Along OX	Along OY
7 kN force	$-7 \cos 40° = -5.36$	$7 \sin 40° = 4.50$
5 kN force	$5 \cos 30° = 4.33$	$5 \sin 30° = 2.50$
8 kN force	$8 \cos 50° = 5.14$	$-8 \sin 50° = -6.13$

ΣF_x = sum of components in direction OX = 4.11 kN
ΣF_y = sum of components in direction OY = 0.87 kN

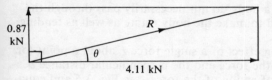

Fig. 8.6 Worked example 8.1

Then $R^2 = 4.11^2 + 0.87^2$
or $R = 4.20$ kN
$\tan \theta = 0.87/4.11 = 0.212$
or $\theta = 12.0°$

The resultant is thus 4.20 kN acting at 12.0° above OX.

8.5 Equilibrium and equilibriant

In Fig. 8.5 the line OC represents the resultant force. If another force represented by the line CO was added to the three already shown in Fig. 8.5, the force diagram would then become a closed polygon. This would mean that the resultant force was zero. The forces are then said to be in equilibrium. The single force which, acting with a number of forces produces a zero resultant force, that is equilibrium, is called the equilibriant. Note that equilibriant and resultant are equal and opposite.

8.6 Conditions for the equilibrium of coplanar concurrent forces

If a system of coplanar, concurrent (all passing through the same point) forces keeps a body at rest, that is in equilibrium, then the condition that the system must satisfy is that the algebraic sums of the components of all the forces in any two directions at right angles must both be zero. The two directions chosen are often the horizontal and the vertical. The condition then becomes

$$\left.\begin{array}{l} \Sigma F_H = 0 \\ \text{and } \Sigma F_V = 0 \end{array}\right\} \qquad [8.1]$$

Here, F_H and F_V represent the horizontal and vertical components of any force F in the horizontal and vertical directions respectively.

8.7 Moment of a force

Coplanar forces acting on a body do not necessarily pass through the same point and so can tend to make the body rotate as well as tending to make it move along.

The moment or turning effect of a single force F about a point (or fulcrum) is the product of the force and the perpendicular distance d from the point to the line of action of the force—see Fig. 3.5 and equation [3.11].

8.8 General conditions for the equilibrium of coplanar forces

If a system of coplanar forces keeps a body in equilibrium, then the conditions that the system must satisfy are those given by [8.1], plus the condition that the algebraic sum of the moments of all the forces about any point in their plane must also be zero,

$$\left.\begin{array}{rl} \text{or} & \Sigma F_H = 0 \\ & \Sigma F_V = 0 \\ \text{and} & \Sigma M = 0 \end{array}\right\} \qquad [8.2]$$

where M represents the moment of a force about any point in the plane of the forces.

If the system has three forces and equilibrium exists, the three forces must either be concurrent or parallel.

Worked example 8.2
A horizontal beam of negligible mass rests on a support at each end. It is loaded as shown in Fig. 8.7. Find the forces (or reactions) at the supports.

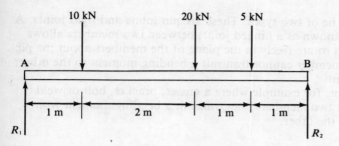

Fig. 8.7 Worked example 8.2

Solution
Let the reactions at the supports be R_1 and R_2. If the beam is in equilibrium $\Sigma F_V = 0$. Thus $(R_1 + R_2)$ must be $(10 + 20 + 5)$, that is, 35 kN.

Also the algebraic sum of the moments of all forces about any point must be zero.

Take moments about A

$5 R_2 = 1 \times 10 + 3 \times 20 + 4 \times 5$
$ = 90$

Thus $R_2 = 18$ kN

Take moments about B

$5 R_1 = 10 \times 4 + 20 \times 2 + 5 \times 1$
$ = 85$

Thus $R_1 = 17$ kN
Check: $R_1 + R_2 = 35$ kN

8.9 Triangle and polygon of forces

If a body is kept in equilibrium by three coplanar forces, then these forces may be represented by the three sides of a triangle taken in order. This triangle is called the triangle of forces (or force triangle).

If a body is kept in equilibrium by more than three forces, then these forces may be represented by the sides of a polygon taken in order. This polygon is called the polygon of forces (or force polygon).

8.10 Framed structures – types of joint

A framed structure consists of a number of members joined together.

The joints may be of two types. These are pin joints and rigid joints. A pin joint (also known as a hinged joint) between two members allows each member to rotate freely in the plane of the members about the pin or hinge. One member cannot transmit a bending moment to the other through the joint.

A rigid joint, for example where a gusset, bracket, bolt or weld is used to connect two members, can transmit a bending moment from one member to the other.

8.11 Types of support

The loads on a framework must be transmitted to supports which provide the reactions necessary to maintain equilibrium. These supports may be of several different types.

A support could be a pin joint rigidly fixed. Such a support can provide both a horizontal reaction H and a vertical reaction V – Fig. 8.8(a).

A support could also be a pin joint on a roller free to move horizontally. This type of support can provide a vertical reaction V only – Fig. 8.8(b).

Fig. 8.8 Types of support

8.12 Static determinancy

Frameworks are conveniently classified as being either statically determinate or statically indeterminate.

A statically determinate plane frame (that is, one with all the members lying in the same plane) is one in which the forces in all the members may be found both in magnitude and in direction by applying the three conditions for the equilibrium of coplanar forces – [8.2].

A statically indeterminate plane frame cannot be analysed using only the conditions for the equilibrium of coplanar forces. This is because the number of unknowns exceeds the number of quantities which may be determined using these conditions. The analysis of statically indeterminate frames requires a knowledge of the cross-sections and second moments of areas of the members.

8.13 Pin-jointed plane frameworks

In order that a plane framework may be statically determinate, several assumptions must be made. These are as follows:

1. All the joints are at the ends of members and are pin joints.
2. The members of the frame are all straight, have negligible weight and do not bend or change in length when the frame is loaded.
3. All loads and support reactions are applied at joints.
4. All members, applied loads and reactions lie in the same plane.

8.14 Conditions for statically determinate frames

If a framework has just enough members to make it stable and statically determinate, then it is called a perfect frame. A simple perfect frame is shown in Fig. 8.9 with a pin at one support and a roller at the other.

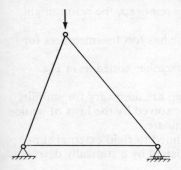

Fig. 8.9 Simple perfect frame

This frame is statically determinate since the forces in all the members can be found by the application of [8.2]. If j represents the number of joints in the frame and m the number of members, then in this case $j = 3$ and $m = 3$.

If the frame in Fig. 8.9 is now extended by adding another triangle as shown in Fig. 8.10, then the resulting frame is still perfect and statically determinate. In this case $j = 4$ and $m = 5$.

If the frame is again extended by adding another triangle (see Fig. 8.11) then $j = 5$ and $m = 7$.

The frame remains statically determinate provided that only triangles are added.

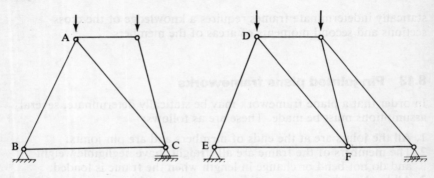

Fig. 8.10   **Fig. 8.11**

Examination of the values of j and m for Figs. 8.9, 8.10 and 8.11 reveals that the relationship between them can be expressed by

$$m = 2j - 3 \qquad [8.3]$$

Equation [8.3] is the condition which must be satisfied by a statically determinate plane framework with a pin joint at one support and a roller at the other.

If the frame in Fig. 8.10 had a member removed, the result might be as in Fig. 8.12.

Here $m = 4$ and $2j - 3 = 5$. This frame has too few members for stability and is thus unstable or substatic.

If the frame in Fig. 8.10 had an extra member added as in Fig. 8.13 then $m = 6$ and $2j - 3 = 5$.

This frame contains more members than are necessary for stability and it is said to be hyperstatic. It cannot be solved by the laws of static equilibrium and hence is statically indeterminate.

If two or more joints (or nodes) are pinned to rigid external supports then the condition which must be satisfied by a statically determinate plane framework becomes

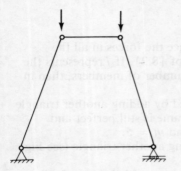

Fig. 8.12 Unstable frame

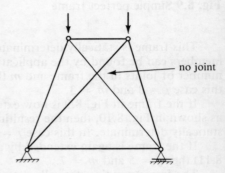

no joint

Fig. 8.13 Hyperstatic frame

$$m = 2(j-s) \qquad [8.4]$$

where j is the total number of joints and s is the number of joints pinned to rigid external supports. A simple example of such a framework is shown in Fig. 8.14.

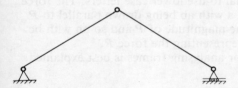

Fig. 8.14

Here, $m = 2$, $j = 3$ and $s = 2$ and hence $m = 2(j-s)$.

8.15 Analysis of statically determinate pin-jointed frameworks

The magnitudes of the forces in the members of a frame can be determined using the conditions for equilibrium as stated in [8.2], namely

$\Sigma F_H = 0$
$\Sigma F_V = 0$
$\Sigma M = 0$

Two methods of analysis will be considered.

8.16 Force diagram method

In this method, the magnitudes of the forces in all the members and reactions are all found by scale drawing. This method uses Bow's notation to identify forces. This can be illustrated by referring to Fig. 8.15.

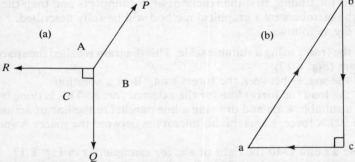

Fig. 8.15 Bow's notation (a) Space diagram (b) Force diagram

Suppose that the three forces P, Q and R are in equilibrium. Using Bow's notation the spaces between the forces are labelled with a capital letter (or number). It is usual to work clockwise round the system. Any force is then represented by the two letters on either side of it, for example, the force P becomes the force AB, Q becomes the force BC, etc. In the force diagram it is usual to use lower case letters. The force AB is drawn starting at the point a with ab being drawn parallel to P and having length representing the magnitude of P and so on with bc representing the force Q and ca representing the force R.

The force diagram method for analysing frames is best explained using worked examples.

Worked example 8.3
Find the reactions at the supports and the magnitude of the force in each member of the pin-jointed frame in Fig. 8.16. State whether each members is a strut or a tie.

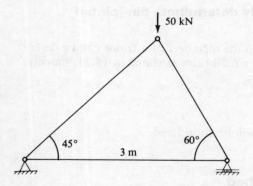

Fig. 8.16 Worked example 8.3

Solution
The method for finding first the reactions at the supports and then the forces in the members by a graphical method will be fully described. The steps are as follows.

1. Draw the frame using a suitable scale. This diagram is called the **space diagram** (Fig. 8.17).
2. Label the spaces between the forces using Bow's notation.
3. Draw the **load** (or **force**) **line** for the external forces. This is done by using a suitable scale and drawing a line parallel to the line of action of the 50 kN force. Label this ab since it lies between the spaces A and B.
4. Choose a point O to the right of ab, for example, as in Fig. 8.17. Draw the lines oa and ob. This diagram is called the **polar** diagram.

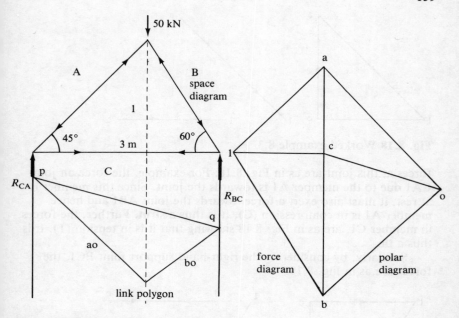

Fig. 8.17 Worked example 8.3

5. Starting at any point p on the line of action of R_{CA}, draw a line ao parallel to ao in the polar diagram. At the point where ao intersects the line of action of the 50 kN force, draw bo parallel to bo in the polar diagram and intersecting the line of action of R_{BC} at q. Join pq. This diagram is called the **link** (or **funicular**) **polygon**.
6. Transfer the line pq to the polar diagram by drawing co parallel to pq.
7. The reactions R_{CA} and R_{BC} are now represented by ac and cb respectively. These are 18.3 kN and 31.7 kN respectively.
8. Draw the **force diagram**. Start, for example, at the left-hand support by drawing the line a1 parallel to the member A1 through a and the line c1 parallel to the member C1 through c. This locates the point 1. Now for the right-hand support, draw a line b1 through b parallel to the member B1 through b. This should pass through the point 1.
9. Use the force diagram to scale off the forces in the members. These are

    F_{A1}, represented by a1, equal to 25.9 kN
    F_{B1}, represented by b1, equal to 36.6 kN
    F_{C1}, represented by c1, equal to 18.3 kN

10. Determine whether the members are in compression or tension as follows. The force polygon for the left-hand support joint CA1 is as in Fig. 8.18.

Since R_{CA}, represented by ca, acts upwards, the directions of the other

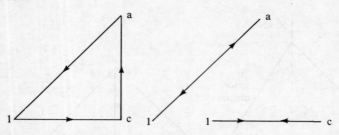

Fig. 8.18 Worked example 8.3

forces at this joint are as in Fig. 8.18. For example, the force on joint CA1 due to the member A1 is towards the joint. Since this member is at rest, it must also exert a force towards the joint AB1 and hence member A1 is in compression (C). It is thus a **strut**. Further, the forces in member C1, are as in Fig. 8.18 showing that it is in tension (T). It is thus a **tie**.

Similarly, by considering the right-hand support joint BC1, the forces are as in Fig. 8.19.

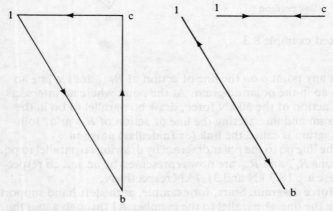

Fig. 8.19 Worked example 8.3

The member B1 is thus in compression and is a **strut**.

The force arrows may all be transferred to the space diagram as shown. The answers are summarized below.

R_{CA}	18.3 kN		
R_{BC}	31.7 kN		
A1	25.9 kN	C	STRUT
B1	36.6 kN	C	STRUT
C1	18.3 kN	T	TIE

Worked example 8.4

Find the magnitude and type of the forces in the members of the pin-jointed framework shown in Fig. 8.20. The length of all the members is 4 m.

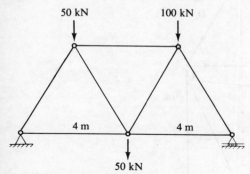

Fig. 8.20 Worked example 8.4

Solution

The steps are as follows:

1. Draw the space diagram to a suitable scale.
2. Label the spaces between the external forces using the letters A, B, C, D and E and the internal spaces 1, 2, 3 and 4. Note that the line of action of the 50 kN load mid-way between the supports has been produced upwards.
3. Draw the load line abcd using a suitable scale with ab representing the force AB (50 kN), bc representing the force BC (100 kN), and cd representing the force CD (50 kN).
4. Choose a suitable pole O and draw the polar diagram by drawing ao, bo, co and do (Fig. 8.21).
5. Construct the link polyon commencing with ao from p parallel to ao in the polar diagram and ending at q with do parallel to do in the polar diagram.
6. Join pq and draw oe parallel to pq.
7. Read off the reactions R_{EA} = ae and R_{DE} = ed from the load line.
8. Draw the force diagram as follows. By considering the joint EA1, locate the point 1 by drawing a1 parallel to member A1 through a and e1 parallel to the member E1 through e. Now locate the point 4 for the joint DE4 by drawing d4 parallel to the member D4 through d and e4 parallel to member E4 through e. Next locate the point 2 for the joint AB21 by drawing b2 parallel to the member B2 through b and 12 parallel to member 12 through 1. Point 3 for joint CD43 can then be located by drawing c3 parallel to the member C3 through C and 43 parallel to member 43 through 4. Finally, complete the force polygon e1234 for the joint E1234 by drawing a line parallel to force 23 through 2. This should pass through 3.

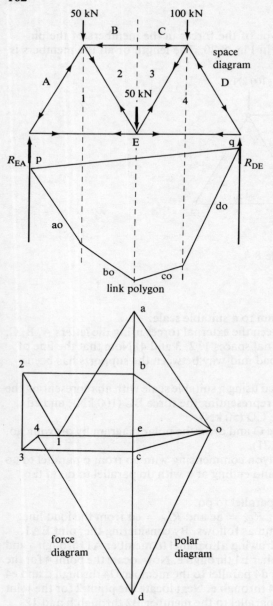

Fig. 8.21 Worked example 8.4

9. Scale off the forces in all the members.
10. Find the type of the force in each member as shown in Fig. 8.22 and mark the force directions on the space diagram.

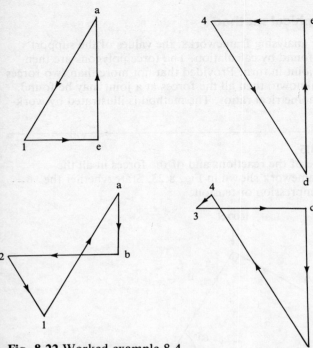

Fig. 8.22 Worked example 8.4

In the force polygon for joint EA1, force ea is known to act upwards.
Member E1 is thus in tension and member A1 in compression.
In the force polygon for joint DE4, force de is known to act upwards.
Member E4 is thus in tension and member D4 in compression.
In the force polygon for joint AB21, force ab is known to be downwards. Member B2 is thus in compression and member 12 in tension.
In the force polygon for joint CD43, force CD is known to be downwards. Member C3 is thus in compression — C3 and B2 are parts of the same member — and member 34 is in tension.

The answers are summarized below.

R_{EA}	88 kN	
R_{DE}	112 kN	
A1	101 kN	C
B2 or C3	72 kN	C
D4	130 kN	C
12	43 kN	T
34	14 kN	T
E1	51 kN	T
E4	65 kN	T

8.17 Semi-graphical method

In this method for analysing frameworks, the values of the support reactions are first found by calculation. The force polygons are then sketched for each joint in turn. Provided that not more than two forces at any joint are unknown, then all the forces at a joint may be found using simple trigonometrical ratios. The method is illustrated by worked examples.

Worked example 8.5
Find the magnitude of the reactions and of the forces in all the members of the framework shown in Fig. 8.23. State whether the members are in compression or tension.

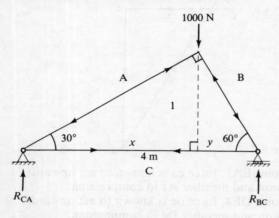

Fig. 8.23 Worked example 8.5

Solution
The steps are as outlined below.

1. Label the spaces between forces using Bow's Notation.
2. Find the reactions by taking moments, first about the left-hand support joint and then about the right-hand support joint. This gives

   $4 R_{BC} = 1000x$
   and $4 R_{CA} = 1000y$

   Now since length of member C1 is 4 m and the frame is a 30°, 60°, 90° triangle, then B1 is 2 m long and hence $y = 1$ m and $x = 3$ m.

   Thus $4 R_{BC} = 3000$
   and $4 R_{CA} = 1000$
   Thus $R_{BC} = 750$ N
   and $R_{CA} = 250$ N
   Check: $R_{BC} + R_{CA} = 1000$ N = total load

3. Sketch the force polygon for the joint CA1 (Fig. 8.24). The directions of the forces in the members must be as shown to maintain equilibrium of the joint.

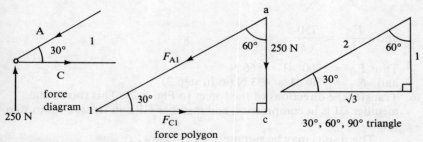

Fig. 8.24 Worked example 8.5

By comparing the force polygon with a 30°, 60°, 90° triangle, it should be clear that

$$\frac{F_{A1}}{2} = \frac{250}{1}$$

or $F_{A1} = 500$ N

and $\frac{F_{C1}}{\sqrt{3}} = \frac{250}{1}$

or $F_{C1} = 250\sqrt{3} = 433$ N

4. Transfer the directions of the forces to Fig. 8.23. This shows that member A1 is in compression and that member C1 is in tension.
5. Repeat steps 3 and 4 for the joint BC1 (Fig. 8.25).

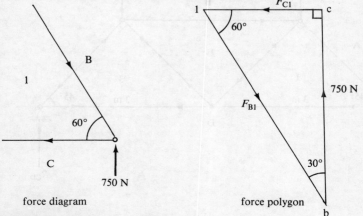

Fig. 8.25 Worked example 8.5

From Fig. 8.25

$$\frac{F_{B1}}{2} = \frac{750}{\sqrt{3}}$$

and $\quad \dfrac{F_{C1}}{1} = \dfrac{750}{\sqrt{3}}$

Thus $F_{B1} = 1500/\sqrt{3} = 866$ N
and $\ F_{C1} = 750/\sqrt{3} = 433$ N (as in step 3)

6. Transfer the directions of the forces to Fig. 8.23. This shows that member B1 is in compression and member C1 is in tension (as in step 4).

The results may be summarized as below.

R_{CA}	250 N	
R_{BC}	750 N	
A1	500 N	C
B1	866 N	C
C1	433 N	T

Worked example 8.6
A loaded framework is shown in Fig. 8.26. Calculate the reactions at the supports and find the forces in all the members. Indicate which members are struts and which are ties.

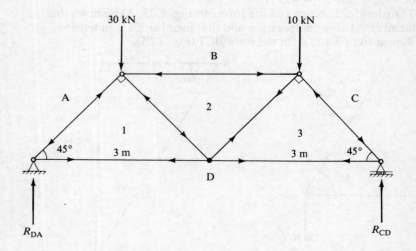

Fig. 8.26 Worked example 8.6

Solution

1. Label spaces between forces using Bow's notation.
2. Take moments about joint DA1 and about joint CD3. Since the triangles are all right-angled, the line of action of the 30 kN is 1.5 m from joint DA1 and the line of action of the 10 kN is 1.5 m from joint CD3.

Thus $6 R_{CD} = 30 \times 1.5 + 10 \times 4.5$
or $\quad R_{CD} = 15$ kN
and $6 R_{DA} = 30 \times 4.5 + 10 \times 1.5$
or $\quad R_{DA} = 25$ kN
Check: $R_{DA} + R_{CD} = 40$ kN = total load

3. Draw the force polygon for the joint DA1 (Fig. 8.27).

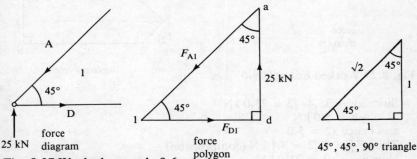

Fig. 8.27 Worked example 8.6

$$\frac{F_{A1}}{\sqrt{2}} = \frac{25}{1}$$

and $\quad \dfrac{F_{D1}}{1} = \dfrac{25}{1}$

Thus $F_{A1} = 25 \sqrt{2} = 35.4$ kN (compression)
and $\quad F_{D1} = 25.0$ kN (tension)

Member A1 is in compression and member D1 is in tension.

4. Draw the force polygon for the joint CD3 (Fig. 8.28).

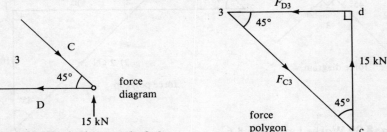

Fig. 8.28 Worked example 8.6

$F_{D3} = 15.0$ kN (tension)
$F_{C3} = 15\sqrt{2} = 21.2$ kN (compression)

Member C3 is in compression and member D3 is in tension.
5. Draw the force polygon for the joint AB21 (Fig. 8.29).

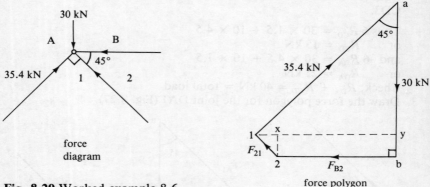

Fig. 8.29 Worked example 8.6

Since $ay = 35.4/\sqrt{2} = 25.0$ kN
$yb = 5.0$ kN
and hence $x2 = 5.0$
Thus $F_{21} = 5\sqrt{2} = 7.1$ kN (compression)
Now $ay = y1 = 25.0$ kN
Thus since $x2 = x1 = 5.0$
$xy = 20.0$
Thus $F_{B2} = 20.0$ kN (compression)

Member B2 is in compression and member 21 is in compression.
6. Draw the force polygon for the joint BC32 (Fig. 8.30).

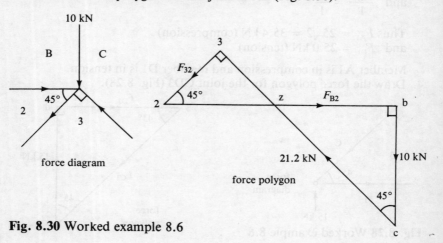

Fig. 8.30 Worked example 8.6

Since cz = 10 $\sqrt{2}$ = 14.1
   z3 = 7.1
Since z3 = 23
F_{32} = 7.1 kN (tension)
Also z2 = 7.1 $\sqrt{2}$ = 10.0 and zb = bc = 10 kN
and hence b2 = 20.0
or F_{B2} = 20.0 kN (agreeing with step 5)

Member 23 is in tension.
7. Transfer the directions of all the forces to Fig. 8.26.
   Summary:

R_{DA}	25 kN	
R_{CD}	15 kN	
A1	35.4 kN	Strut
B2	20.0 kN	Strut
C3	21.2 kN	Strut
D1	25.0 kN	Tie
D3	15.0 kN	Tie
21	7.1 kN	Strut
32	7.1 kN	Tie

8.18 Cantilever frameworks

The methods used to analyse the simple frames in the previous sections can be applied to statically determinate cantilever frames. When these are analysed using the force diagram method it is not necessary, in many cases, to determine the reactions at the supports before drawing the force diagram. The worked examples following show how cantilever frames may be analysed, both by the force diagram method and by the semi-graphical method.

Worked example 8.7
Find the forces in all the members of the framework shown in Fig. 8.31 and indicate which members are in tension and which are in compression.

Solution
The steps in the solution are as follows.
1. Draw the space diagram to a suitable scale.
2. Label the spaces between the forces. (Fig. 8.32) The reaction R_1 will be equal and opposite to the internal force in the member AB. Reaction R_2 must be the equilibrant of the internal forces in the members A1 and D1.
3. Draw the load line bcd using a suitable scale.

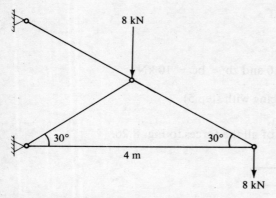

Fig. 8.31 Worked example 8.7

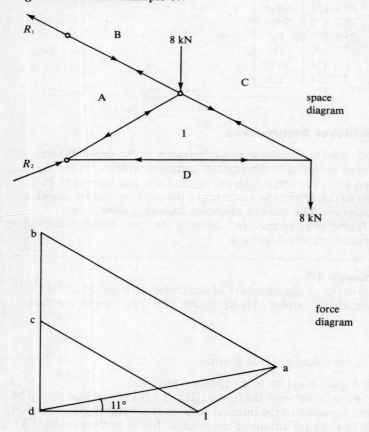

Fig. 8.32 Worked example 8.7

4. Draw the force diagram, starting with the joint CD1.
5. Scale off the values of the forces in the members and the reactions.
6. Measure the angle made by R_2 with the horizontal.
7. Determine the directions of the forces at the joints (see worked example 8.3 step 10) and mark them on the space diagram. The results are summarized below.

R_1	24.0 kN, 30° to horizontal	
R_2	21.2 kN, 11° to horizontal	
AB	24.0 kN	T
A1	8.0 kN	C
C1	16.0 kN	T
D1	13.9 kN	C

Worked example 8.8
Find the magnitude and type of the forces in the members of the pin-jointed cantilever framework shown in Fig. 8.33 and the reactions at the wall.

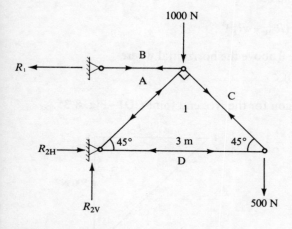

Fig. 8.33 Worked example 8.8

Solution
1. Label the spaces between the forces.
2. The reaction R_1 must be equal and opposite to the internal force exerted on the joint by the member AB. Let the horizontal and vertical components of the reaction R_2 at the lower support be R_{2H} and R_{2V} respectively as shown in Fig. 8.33. Taking moments about the lower

joint DA1 gives

$1.5 R_1 = 1.5 \times 1000 + 3 \times 500$
or $R_1 = 2000$ N

Since R_1 and R_{2H} are the only horizontal forces acting
$R_{2H} = R_1 = 2000$ N

Using $\Sigma F_V = 0$
$1000 + 500 - R_{2V} = 0$
or $R_{2V} = 1500$ N

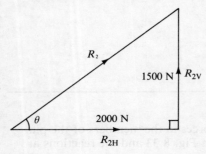

Fig. 8.34 Worked example 8.8

From Fig. 8.34 $R_2 = (R_{2H}^2 + R_{2V}^2)^{\frac{1}{2}}$
thus $R_2 = 2500$ N
R_2 will act at an angle θ above the horizontal where
$\theta = \tan^{-1}(1500/2000)$
$= 36.9°$
3. Sketch the force polygon for the free end joint CD1–Fig. 8.35.

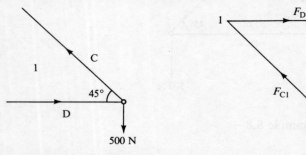

Fig. 8.35 Worked example 8.8

From the force polygon
$F_{D1} = 500$ N
and $F_{C1} = 500/\sin 45° = 707$ N
4. Transfer the directions of the forces to Fig. 8.33.

5. Sketch the force polygon for the joint ABC1 – Fig. 8.36.

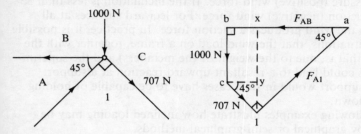

Fig. 8.36 Worked example 8.8

y1 = 707 sin 45°
   = 500
∴ x1 = 1500

Thus F_{A1} = 1500/sin 45°
or F_{A1} = 2121 N
ax = x1 = 1500
cy = y1 = 500

Thus F_{AB} = 2000 N, agreeing with the value found above for R_1.
6. Transfer the directions of the forces to Fig. 8.33. The results are summarized below.

R_1	2000 N, horizontal	
R_2	2500 N, 36.9° to horizontal	
AB	2000 N	T
A1	2121 N	C
C1	707 N	T
D1	500 N	C

8.19 Inclined loading

All the examples dealt with so far have involved vertical loading only. With horizontal or inclined loading, one or more support joints must be capable of providing the necessary horizontal component of reaction. The pressure due to the wind blowing onto a surface is assumed to act normally to the surface. This can produce inclined loading, on a roof truss for example. The direction of the wind pressure depends on the angle between the direction of the wind and the surface. For a horizontal wind the following general rules may be used. For windward surfaces inclined at greater than 35° to the horizontal the wind produces a pressure (positive) wind force. If the inclination is less than 35° there is a suction (negative) wind force. For leeward surfaces at all inclinations the wind produces a suction force. In practice, it is possible in certain situations, that the wind load on a frame, together with the dead load (that is, due to the weight of the members) and the superimposed loads could lead to a resultant upward reaction at a support joint. The support would in such cases have to be capable of holding the frame down.

The following examples illustrate how inclined loading may be dealt with by graphical or semi-graphical methods.

Worked example 8.9
The roof truss shown in Fig. 8.37 has a pin joint at one support and a roller at the other. Find the reactions at the supports and the forces in all the members of the truss. Indicate which members are ties and which are struts.

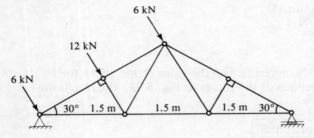

Fig. 8.37 Worked example 8.9

Solution
The problem can be solved graphically as follows.

1. Draw the space diagram to a suitable scale.
2. Label the spaces – Fig. 8.38.
3. Since the inclined loading is symmetrical about the 12 kN load, the resultant inclined load, namely 24 kN, must act along the same line as the 12 kN load. The frame is thus kept in equilibrium by three forces, the resultant inclined load, R_{EA} and R_{DE}. These must all pass through the same point. This point can be found by finding where the line of the 12 kN load and the line of R_{DE} meet. This is at d. (The reaction R_{DE} must be vertical since the joint DE5 is on a roller.) The line of R_{EA} ca then be drawn through d.
4. Using a suitable scale, mark off the line dcba with cd, bc and ab representing 6 kN, 12 kN and 6 kN respectively.
5. Through a draw ae parallel to R_{EA}. The line ea then represents R_{EA} and de represents R_{DE}.

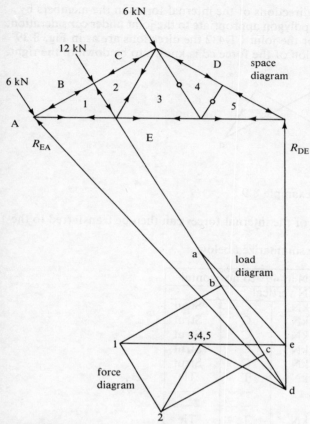

Fig. 8.38 Worked example 8.9

6. For the joint DE5, locate the point 5 by drawing e5 through e parallel to member E5 and d5 through d parallel to member D5.
7. For the joint D54, locate the point 4 by drawing 45 through 5 parallel to member 45 and d4 through d parallel to member D4. This shows that points 4 and 5 are the same point.
8. For the joint E345, locate the point 3 by drawing e3 through e parallel to member E5 and 34 through 4 parallel to member 34. This shows that the points 3, 4 and 5 are all the same point.
9. For the joint EAB1, locate the point 1 by drawing b1 through b parallel to member B1 and e1 through 1 parallel to member E1.
10. For the joint BC21, locate the point 2 by drawing c2 through c parallel to member C2 and 12 through 1 parallel to member 12.
11. For the joint CD432 draw a line through 3 parallel to member 23. This should pass through 2, thus completing the force diagram.
12. Scale off all the forces from the force diagram.

13. Determine the directions of the internal forces in the members by using the force polygon appropriate to the joint under consideration. For example for the joint CD432 the directions are as in Fig. 8.39 since the direction of the force cd is known to be down to the right.

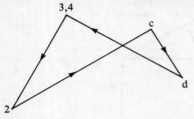

Fig. 8.39 Worked example 8.9

The directions of the internal forces can then be transferred to the force diagram.
The results are summarized below.

R_{EA}	18.3 kN at 49° to horizontal		
R_{DE}	6.9 kN vertical		
B1	17.3 kN	C	Strut
C2	17.3 kN	C	Strut
D4	13.8 kN	C	Strut
D5	13.8 kN	C	Strut
12	12 kN	C	Strut
23	12 kN	T	Tie
34	0	–	–
45	0	–	–
E1	24 kN	T	Tie
E3	12 kN	T	Tie
E5	12 kN	T	Tie

Note that for this loading arrangement members 34 and 45 are redundant. They could be removed without changing the equilibrium of the system or the forces in the other members.

Worked example 8.10
Find the reactions at the supports of the loaded frame shown in Fig. 8.40 and the forces in the members. Which members are in compression and which are in tension?

Solution
Since there is a roller at the joint FA1, the reaction there must be vertical. The joint DEF4 must provide the horizontal reaction to counteract the horizontal components of the inclined loading.

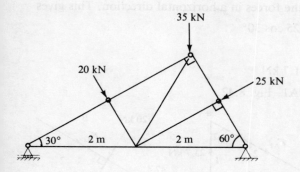

Fig. 8.40 Worked example 8.10

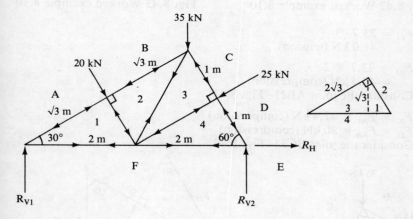

Fig. 8.41 Worked example 8.10

Let the horizontal and vertical components of the right hand reaction be R_H and R_{V2} respectively.

1. Label the spaces between the forces.
2. Take moments about the support DEF4

    $R_{V1} \times 4 = 20 \times \sqrt{3} + 25 \times 1 + 35 \times 1$
    or $R_{V1} = 23.7$ kN

3. Take moments about the support FA1

    $R_{V2} \times 4 = 25 \times 1 + 35 \times 3 + 20 \times \sqrt{3}$
    or $R_{V2} = 41.2$ kN

4. To find R_H resolve the forces in a horizontal direction. This gives

$$R_H + 20 \cos 60° = 25 \cos 30°$$

That is

$$R_H = 21.7 - 10 = 11.7 \text{ kN}$$

5. Consider the joint FA1 – Fig. 8.42.

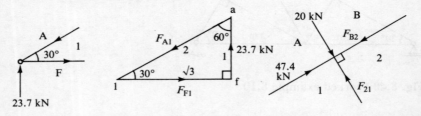

Fig. 8.42 Worked example 8.10 **Fig. 8.43** Worked example 8.10

$$F_{F1} = 23.7 \sqrt{3}$$
$$= 41.0 \text{ kN (tension)}$$

$$F_{A1} = 23.7 \times 2$$
$$= 47.4 \text{ kN (compression)}$$

6. Consider the joint AB21 – Fig. 8.43.

$$F_{B2} = F_{A1} = 47.4 \text{ kN (compression)}$$
$$F_{21} = F_{AB} = 20 \text{ kN (compression)}$$

7. Consider the joint BC32 – Fig. 8.44.

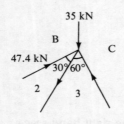

Fig. 8.44 Worked example 8.10

$$bq = 35 \times \tfrac{1}{2} = 17.5$$
Thus $q2 = 47.4 - 17.5 = 29.9$
Thus $F_{23} = 29.9 \times 2/\sqrt{3} = 34.5$ kN (tension)
Also $q3 = 34.5 \times \tfrac{1}{2} = 17.3$
and $qc = 35 \times \sqrt{3}/2 = 30.3$

Thus $F_{C3} = 30.3 + 17.3$
$= 47.6$ kN (compression)

8. Consider the joint CD43 – Fig. 8.45.

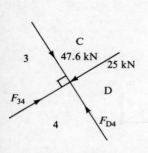
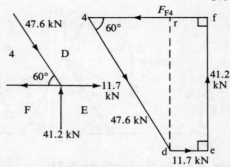

Fig. 8.45 Worked example 8.10 **Fig. 8.46** Worked example 8.10

$F_{D4} = 47.6$ kN (compression)
$F_{34} = 25$ kN (compression)

9. Consider the joint DEF4 – Fig. 8.46.

   r4 = $47.6 \times \frac{1}{2} = 23.8$
   Thus $F_{F4} = 23.8 + 11.7 = 35.5$ kN (tension)

The directions can now be transferred to the space diagram – Fig. 8.41. The results are summarized below, (C = Compression. T = Tension)

R_{V1}	23.7 kN	
R_{V2}	41.2 kN	
R_H	11.7 kN	
A1	47.4 kN	C
B2	47.4 kN	C
C3	47.6 kN	C
D4	47.6 kN	C
F1	41.0 kN	T
F4	35.5 kN	T
21	20.0 kN	C
23	34.5 kN	T
34	25.0 kN	C

8.20 Combined vertical and inclined loading

The example below illustrates the method used to deal with a frame subjected to vertical loading plus a positive wind load. A similar method could be used if the wind load was negative.

Worked example 8.11
Find, by a graphical method, the reactions and the forces in all the members of the frame in Fig. 8.47, and state which members are struts and which are ties.

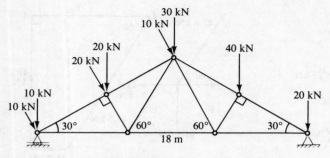

Fig. 8.47 Worked example 8.11

Solution
1. Draw the space diagram to a suitable scale. Reaction R_{GA} will be vertical and the direction of R_{FG} is unknown.
2. Combine the vertical and inclined loads using the parallelogram rule and label the spaces—Fig. 8.48.
3. Commence drawing the external load polygon by drawing the line abcdef with its sections parallel to the appropriate forces.
4. Choose a suitable point O and draw the polar diagram.
5. Construct the link polygon. Since the direction of R_{FG} is unknown, the only point on R_{FG} known is the support joint itself and hence the drawing of the polygon must commence at this point.
6. Transfer the closing side og of the link polygon to the polar diagram.
7. The reaction R_{GA} is vertical. If a line is drawn vertically through a to represent R_{GA} then the point g can be located as shown in Fig. 8.48. The external force polygon abcdefg is now complete. Lines ga and fg represent R_{GA} and R_{FG} respectively.
8. Draw the force diagram for the frame in the usual way.
9. Scale off and determine the directions of the forces in all the members and mark the directions on the space diagram.
   The results are as summarized below.

R_{GA} = 73.1 kN vertical		
R_{FG} = 83.9 kN at 14° to vertical		
B1	108.9	Strut
C2	98.9	Strut
D4	103.1	Strut
E5	123.1	Strut
12	37.3	Strut
23	37.3	Tie
34	34.6	Tie
45	34.6	Strut
G1	89.3	Tie
G3	52.0	Tie
G5	86.6	Tie

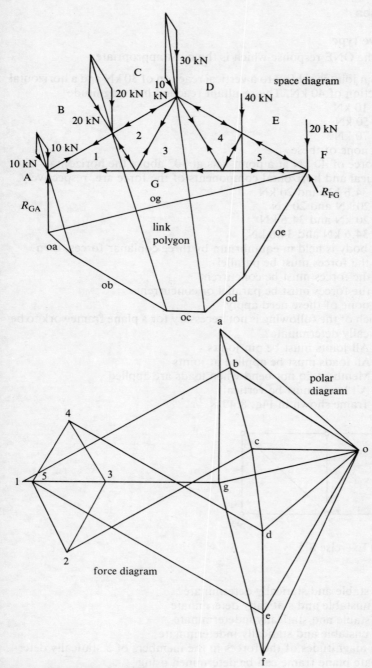

Fig. 8.48 Worked example 8.11

Exercises

Objective type
Choose the ONE response which is the most appropriate.

8.1 A pin joint is subject to a vertical reaction of 30 kN and a horizontal reaction of 40 kN. The resultant reaction has magnitude
    A   10 kN
    B   50 kN
    C   70 kN
    D   none of these

8.2 A force of 40 kN at a point acts at 30° above the horizontal. The vertical and horizontal components of the force are, respectively,
    A   34.6 kN and 20 kN
    B   20 kN and 20 kN
    C   20 kN and 34.6 kN
    D   34.6 kN and 34.6 kN

8.3 If a body is held in equilibrium by three coplanar forces, then
    A   the forces must be parallel
    B   the forces must be concurrent
    C   the forces must be parallel or concurrent
    D   none of these need apply

8.4 Which of the following is not necessary for a plane framework to be statically determinate?
    A   All joints must be pin joints
    B   All loads must be applied at joints
    C   Members do not bend when loads are applied
    D   All loads must be vertical

8.5 The frame shown in Fig. 8.49 is

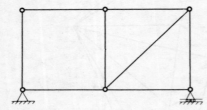

Fig. 8.49 Exercise 8.5

    A   stable and statically determinate
    B   unstable and statically determinate
    C   stable and statically indeterminate
    D   unstable and statically indeterminate

8.6 The magnitudes of the forces in the members of a statically determinate plane frame can be determined using

A $\Sigma F_H = 0$ and $\Sigma F_V = 0$
B $\Sigma M = 0$
C $\Sigma F_H = 0$ (or $\Sigma F_V = 0$) and $\Sigma M = 0$
D $\Sigma F_H = 0$, $\Sigma F_V = 0$ and $\Sigma M = 0$

Problems

8.7 Are the frames shown in Fig. 8.50 statically determinate? If not, suggest simple means by which they could be made statically determinate.

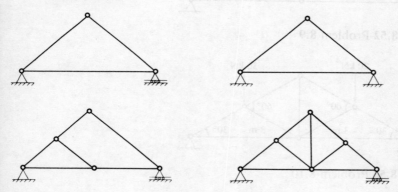

Fig. 8.50 Problem 8.7

8.8 to 8.20 Find, using a graphical method, the reactions and the forces in all the members for the pin-jointed frames in Fig. 8.51 to Fig. 8.63. State whether the members are struts or ties.

8.21 to 8.30 Calculate the reactions at the supports of the frames in Fig. 8.51–8.55 and in Fig. 8.57–8.61 and then find the forces in all the members using force diagrams and trigonometrical ratios. State whether the members are struts or ties.

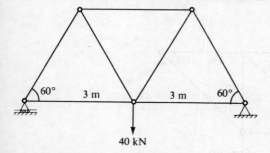

Fig. 8.51 Problem 8.8

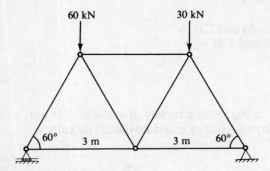

Fig. 8.52 Problem 8.9

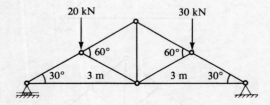

Fig. 8.53 Problem 8.10

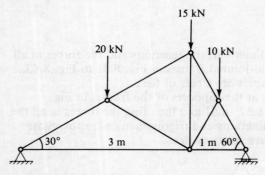

Fig. 8.54 Problem 8.11

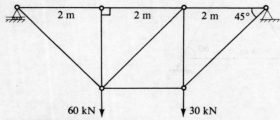

Fig. 8.55 Problem 8.12

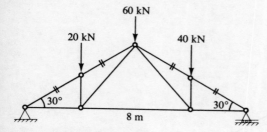

Fig. 8.56 Problem 8.13

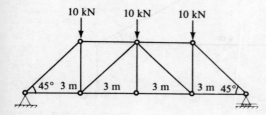

Fig. 8.57 Problem 8.14

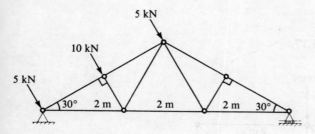

Fig. 8.58 Problem 8.15

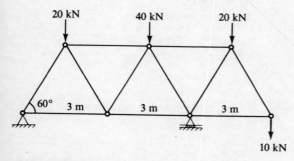

Fig. 8.59 Problem 8.16

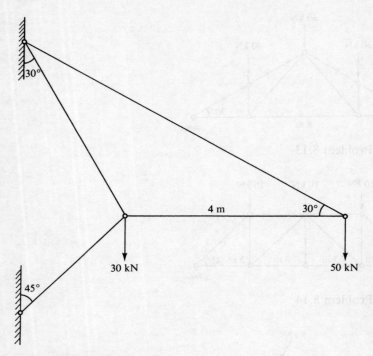

Fig. 8.60 Problem 8.17

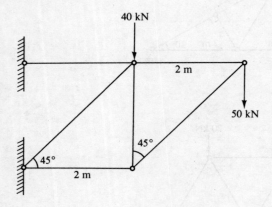

Fig. 8.61 Problem 8.18

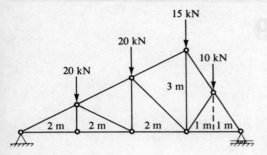

Fig. 8.62 Problem 8.19

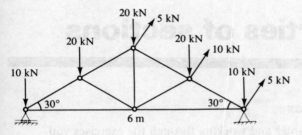

Fig. 8.63 Problem 8.20

Chapter 9

Properties of sections

Learning objectives

After reading this chapter and working through the exercises you should be able to:
- explain, using examples, how bending produces compression and tension across the depth of a beam and that a neutral axis is produced;
- prove, using area moments, that the positions of the neutral axes of a section are dependent on its shape;
- calculate second moments of area about the neutral axes of various cross-sections and relate them to the stiffness of the section;
- determine the section moduli, about major axes, of cross-sections analysed above and relate them to resistance to bending;
- determine the radii of gyration about the major axis for cross-sections analysed above and relate them to resistance to buckling of columns and struts.

Several very important geometrical properties of structural sections depend on the shape of the section. These properties include: Cross-sectional area; Position of centroid; Second moment of area; Section modulus; Radius of gyration.

9.1 Area and centroid

Cross-sectional area needs no explanation. A common unit for

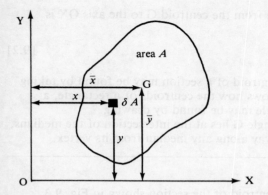

Fig. 9.1 Notation for centroid

measuring area is the mm² since in practice dimensions are often quoted in mm.

The centroid of an area has already been mentioned in Section 5.10.

Consider the small element δA of a plane area A as in Fig. 9.1. The moment of the area of δA about the axis OY is defined as moment of area of δA about OY $= x\delta A$

If the centre of area or centroid G of the whole area A lies a distance $\bar{x}$ from OY, then

$\Sigma x \delta A = A\bar{x}$

or $\quad \bar{x} = \dfrac{\Sigma x \delta A}{A}$ \hfill [9.1]

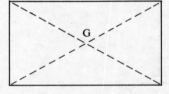

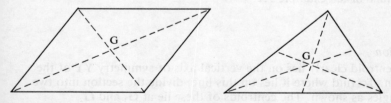

Fig. 9.2 Centroids of rectangle, parallelogram and triangle

Similarly, the distance from the centroid G to the axis OX is

$$\bar{y} = \frac{\Sigma y \delta A}{A} \qquad [9.2]$$

The position of the centroid of a section may be found by taking area moments. Fig. 9.2 shows how the centroids of a rectangle, a parallelogram and a triangle may be found by drawing.

In the case of the triangle G lies at the intersection of the medians, that is, two thirds of the way along any median from its vertex.

Worked example 9.1
Find the position of the centroid of the section shown in Fig. 9.3.

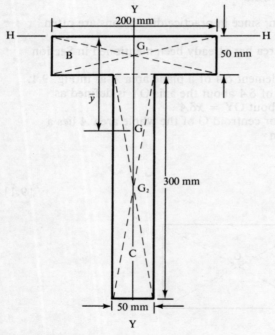

Fig. 9.3 Worked example 9.1

Solution
The centroid clearly lies on the vertical axis of symmetry YY of the section. To find where it lies on this line, divide the section into two rectangles as shown. The centroids of these lie at G_1 and G_2 respectively.

Area of rectangle B = 200 × 50 = 1 × 10⁴ mm²
Area of rectangle C = 300 × 50 = 1.5 × 10⁴ mm²
Total area = 2.5 × 10⁴ mm²

Let the centroid of the section lie at G a distance $\bar{y}$ from the line HH. Taking area moments about HH gives

$1 \times 10^4 \times 25 + 1.5 \times 10^4 \times 200 = 2.5 \times 10^4 \times \bar{y}$
Thus $2.5 \bar{y} = 325$
or $\bar{y} = 130$ mm

The centroid thus lies on the axis of symmetry and 130 mm from HH.

9.2 Neutral plane and neutral axis

Consider the two simple cases shown in Fig. 9.4 of a beam and a cantilever, initially straight and horizontal, loaded as indicated.

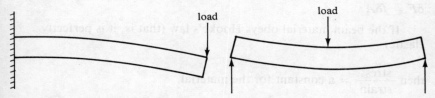

Fig. 9.4 Simply supported beam and simple cantilever

In the case of the beam, loading causes the length of the upper surface to decrease. This means that this surface is in a state of compression. Further, the bottom surface is in a state of tension since the bending produced by the load has caused an increase in length. Clearly, somewhere between the top and bottom surface, the state of compression must change to a state of tension. The layer or plane where there is neither compression or tension is known as the neutral plane. The neutral axis (NA) of a section is the straight line where the neutral plane cuts the section.

In the case of the cantilever, the bending produced by the load causes the upper surface to be in a state of tension and the lower surface to be in a state of compression.

To find where the neutral axis lies, consider for simplicity, a rectangular loaded beam section as in Fig. 9.5.

Consider a strip of area δA, as shown shaded in Fig. 9.5, lying a distance y from NA whose position is as yet not known. If the depth of the strip is sufficiently small, the stress f may be assumed to be the same at all points in it.

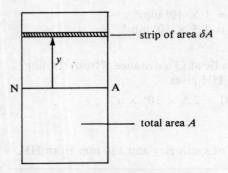

Fig. 9.5 Rectangular beam section

Now

$$\text{Stress} = \frac{\text{load (or force)}}{\text{area}}$$

Thus the force δF due to δA counteracting the bending is

$$\delta F = f\delta A$$

If the beam material obeys Hooke's law (that is, it is perfectly elastic)

then $\dfrac{\text{stress}}{\text{strain}}$ = a constant for the material.

This constant is called Young's modulus. Thus Stress ∝ Strain

where Strain = $\dfrac{\text{change in length}}{\text{original length}}$

Further, if it is assumed that the layers of the beam are bent into the arcs of concentric circles, it is easy to show that the change in length of a layer is directly proportional to y the distance from the NA.

Hence Strain ∝ y and Stress $f \propto y$
and thus force $\delta F \propto y\delta A$
or $\qquad \delta F = ky\delta A \qquad$ [9.3]

where k is a constant depending on the beam material.

The total force F counteracting bending is the sum $\Sigma \delta F$.

Hence $F = \Sigma ky\delta A$

However, for equilibrium, F must be zero, that is the sum of compressive and tensile forces must be zero.

Thus $\Sigma y\delta A$ must be zero. The only way for this condition to be satisfied is for all the y values to be measured from the centroid. This means that

The neutral axis must pass through the centroid of the cross-section.

All the beam layers above the NA will be in compression and all those below will be in tension. For the cantilever the reverse will be true.

The position of the NA together with the variations of stress and strain are shown in Fig. 9.6 for several simply-supported loaded beam sections. In these diagrams AB and CD represent the maximum compressive and tensile stresses respectively. Similarly, EF and HJ represent the maximum compressive and tensile strains respectively.

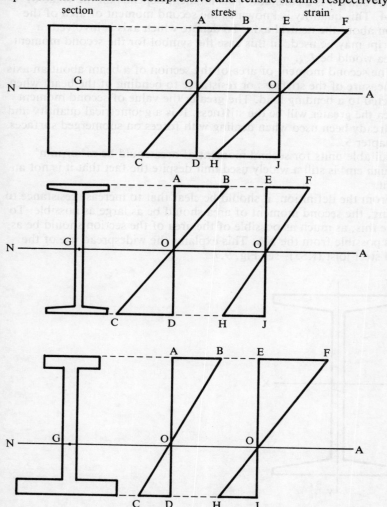

Fig. 9.6 Stress and strain variations for three sections

9.3 Second moment of area

In the previous section the force δF opposing bending of a loaded beam, by [9.3], was

$$\delta F = ky\delta A$$

for an element of area δA of the beam section situated distance y from the NA.

The moment of the force opposing bending is thus $y(ky\delta A)$ or $ky^2\delta A$. For the whole area A the opposing or resisting moment is thus $k\Sigma y^2\delta A$ since k is a constant. This moment depends on the value of $\Sigma y^2\delta A$. This quantity is known as the second moment of area of the section about the neutral axis. To denote which axis is involved, a subscript may be used. In this case the symbol for the second moment of area would be I_{NA}.

The second moment of area of the section of a beam about an axis is a measure of the stiffness or resistance to bending of the beam when subjected to a bending load. The greater the value of second moment of area the greater will be the stiffness. It is a geometrical quantity and has already been used when dealing with forces on submerged surfaces in Chapter 5.

Suitable units for second moment of area could be m^4 or mm^4 although cm^4 is still a widely used unit despite the fact that it is not an SI unit.

From the definition, it should be clear that to increase resistance to bending, the second moment of area should be as large as possible. To ensure this, as much as possible of the area of the section should be as far as possible from the NA. This explains the widespread use of the rolled steel joist (RSJ)—see Fig. 9.7.

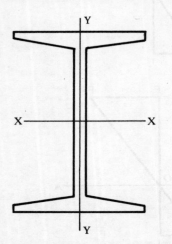

Fig. 9.7 Rolled steel joint section

The value of I_{XX} is much larger than I_{YY} and hence the resistance to bending about XX is much greater than about YY. The axes XX and YY which intersect at the centroid of a section are known as the principal (or major) axes.

9.4 Section modulus

The ratio I/y occurs frequently in problems involving bending stress. Both are geometrical properties of a section and the ratio is known as the section modulus. It is denoted by Z.

Thus $Z = \dfrac{I}{y}$

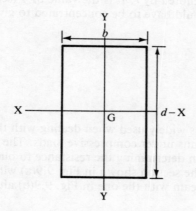

Fig. 9.8 Rectangular section – principal axes

For a rectangular section as in Fig. 9.8, only the section moduli for the layers furthest from the axes where the stresses are maximum are of practical interest.
These are

$$Z_{XX} = \frac{I_{XX}}{d/2}$$

and $Z_{YY} = \dfrac{I_{YY}}{b/2}$ [9.5]

The value of the section modulus for a given axis depends on the value of I for that axis. Since I is a measure of the resistance to bending, so also is Z. Increasing Z increases the resistance to bending. Usually the important value of Z is for that axis for which Z is a maximum, that is, for which the resistance to bending is greatest. For the section in Fig. 9.8 this would be

$$Z_{XX} = \frac{I_{XX}}{d/2}$$

The section modulus may be measured in m³ although mm³ is a more commonly used unit. Even though it is not an SI unit, cm³ is also widely used.

9.5 Radius of gyration

This is another geometrical property of a section which takes into account both the area of cross-section A and second moment of area I. Radis of gyration may be denoted by r. It is the value of y (see Fig. 9.5) at which all the area would have to be concentrated to give the same value of I as
$\Sigma y^2 \delta A$

Thus $I = r^2 A$

or $\quad r = \sqrt{\dfrac{I}{A}}$ [9.6]

The radius of gyration is widely used when dealing with the buckling of struts and columns under compressive loads. The shape of a section is very important in determining the resistance to buckling. For example, a beam with the section shown in Fig. 9.9(a) will buckle much more readily than a beam with the one in Fig. 9.9(b) although the areas are the same.

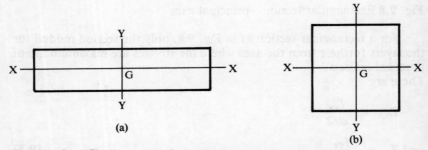

Fig. 9.9 Two sections with the same cross-sectional area A

A column with section as in Fig. 9.9(a) will buckle about the axis XX rather than about the axis YY because the XX axis is the one about which the second moment of area is least. To summarise then, radius of gyration is one factor which determines the resistance to buckling and the important value of r is its value about the axis giving the least second moment of area. For the section in Fig. 9.9(a) this would be

$$r_{XX} = \sqrt{\frac{I_{XX}}{A}}$$ [9.7]

Suitable units in [9.7] could be mm⁴ for I_{XX}, mm² for A and hence mm for r.

9.6 Calculation of values for second moment of area, radius of gyration and section modulus

Only the rectangular, T and I sections will be considered.

Rectangular section

Consider the shaded strip shown in Fig. 9.10. Its area is $b\,dy$ and hence its second moment of area about XX is $by^2\,dy$. For the whole rectangle, then,

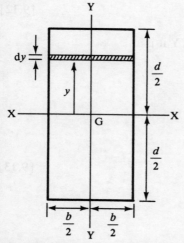

Fig. 9.10 Notation for second moment of area

$$I_{XX} = \int_{-d/2}^{+d/2} by^2\,dy = b\left[\frac{y^3}{3}\right]_{-d/2}^{+d/2}$$

or $I_{XX} = \dfrac{bd^3}{12}$ [9.8]

Similarly, $I_{YY} = \dfrac{db^3}{12}$ [9.9]

Further, $r = \sqrt{\dfrac{I}{A}}$

Thus $r_{XX} = \sqrt{\dfrac{I_{XX}}{A}} = \sqrt{\dfrac{bd^3}{12db}} = \sqrt{\dfrac{d^2}{12}} = \dfrac{d}{\sqrt{12}}$ [9.10]

and similarly $r_{YY} = \dfrac{b}{\sqrt{12}}$ [9.11]

The section modulus for the beam fibres furthest from XX is

$Z_{XX} = \dfrac{I_{XX}}{y}$

$= \dfrac{bd^3}{12(d/2)}$

or $Z_{XX} = \dfrac{bd^2}{6}$ [9.12]

and for the beam fibres furthest from YY it is

$Z_{YY} = \dfrac{I_{YY}}{x}$

$= \dfrac{db^3}{12(b/2)}$

or $Z_{YY} = \dfrac{db^2}{6}$ [9.13]

9.7 Parallel axis theorem

Equations [9.8] and [9.9] apply only to axes through the centroid G of the rectangle. To find I for another axis parallel to either XX or YY the theorem called the parallel axis theorem is used. This states that the second moment of area I_{ZZ} about any axis I_{ZZ} parallel to an axis through the centroid of a section of area A is given by

$I_{ZZ} = I_{CG} + Ac^2$ [9.14]

where I_{CG} is the second moment of area of the section about the axis through G and c is the distance between the two parallel axes – see Fig. 9.11.

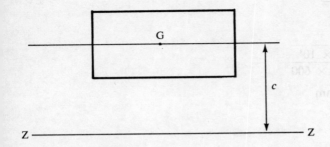

Fig. 9.11 Notation for parallel axis theorem

Worked example 9.2
Find (a) the second moment of area and radius of gyration for the rectangular section shown in Fig. 9.12 about (i) the axis XX, (ii) the axis YY, the value of the section modulus (b) the value of I_{ZZ} and (c) for the beam layer BD.

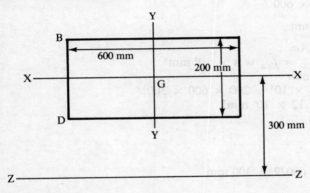

Fig. 9.12 Worked example 9.2

Solution

$$I_{XX} = \frac{bd^3}{12}$$

$$= \frac{600 \times 200^3}{12}$$

$$= 4 \times 10^8 \text{ mm}^4$$

$$r_{XX} = \sqrt{\frac{I_{XX}}{A}}$$

$$= \sqrt{\frac{4 \times 10^8}{200 \times 600}}$$

$$= 57.7 \text{ mm}$$

$$I_{YY} = \frac{db^3}{12}$$

$$= \frac{200 \times 600^3}{12}$$

$$= 3.6 \times 10^9 \text{ mm}^4$$

$$r_{YY} = \sqrt{\frac{I_{YY}}{A}}$$

$$= \sqrt{\frac{3.6 \times 10^9}{200 \times 600}}$$

$$= 173.2 \text{ mm}$$

$I_{ZZ} = I_{CG} + Ac^2$
In this case $I_{CG} = I_{XX} = 4 \times 10^8 \text{ mm}^4$
and $c = 300$ mm
Thus $I_{ZZ} = 4 \times 10^8 + 200 \times 600 \times 300^2$
$= 1.12 \times 10^6 \text{ mm}^4$

$$Z_{YY} = \frac{I_{YY}}{x}$$

long BD $x = 600/2 = 300$ mm

Thus $Z_{YY} = \dfrac{3.6 \times 10^9}{300}$

$$= 1.2 \times 10^7 \text{ mm}^3$$

T-section and I-section
To find the second moment of area for these sections, the parallel axis theorem may be used together with the formula for the second moment of area for a rectangle since both can be divided into rectangles. The centroid of the T-section may be found using the method outlined in section 9.1.

Worked example 9.3
Find the position of the centroid of the beam section shown in Fig. 9.13 and values for

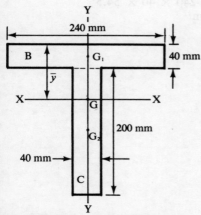

Fig. 9.13 Worked example 9.3

(i) I_{XX}
(ii) I_{YY}
(iii) r_{XX}
(iv) r_{YY}
(v) the section modulus for the beam fibres furthest from XX.

Solution
The centroid may be found by dividing the section into two rectangles B and C and then taking area moments about the line PQ. Let the centroids of B and C be at G_1 and G_2 respectively. This gives

$240 \times 40 \times 20 + 200 \times 40 \times 140 = (240 \times 40 + 200 \times 40)\bar{y}$
Thus $1.92 \times 10^5 + 11.2 \times 10^5 = 1.76 \times 10^4 \bar{y}$
and hence $\bar{y} = 74.5$ mm

Centroid G of the section thus lies on YY 74.5 mm below the top of the section.

(i) For rectangle B
$I_{XX} = I_{CG} + Ac^2$

where $I_{CG} = \dfrac{bd^3}{12}$

$= \dfrac{240 \times 40^3}{12}$

$= 1.28 \times 10^6$ mm^4

and $c = G_1G$
$= 74.5 - 20$
$= 54.5$ mm

Thus for B $I_{XX} = 1.28 \times 10^6 + 240 \times 40 \times 54.5^2$
$= 2.98 \times 10^7$ mm^4

For rectangle C
$I_{XX} = I_{CG} + Ac^2$

where $I_{CG} = \dfrac{bd^3}{12}$

$= \dfrac{40 \times 200^3}{12}$

$= 2.67 \times 10^7$ mm^4

and $c = G_2G$
$= 140 - 74.5$
$= 65.5$ mm

Thus for C $I_{XX} = 2.67 \times 10^7 + 200 \times 40 \times 65.5^2$
$= 6.10 \times 10^7$

Thus total $I_{XX} = 2.98 \times 10^7 + 6.10 \times 10^7$
$= 9.08 \times 10^7$ mm^4

(ii) For rectangle B

$I_{YY} = \dfrac{db^3}{12}$

$= \dfrac{40 \times 240^3}{12}$

$= 4.61 \times 10^7$ mm^4

For rectangle C

$I_{YY} = \dfrac{db^3}{12}$

$= \dfrac{200 \times 40^3}{12}$

$= 1.07 \times 10^6$ mm^4

Thus total $I_{YY} = 4.61 \times 10^7 + 1.07 \times 10^6$
$= 4.72 \times 10^7$ mm^4

(iii) $r_{XX} = \sqrt{\dfrac{I_{XX}}{A}}$

$= \sqrt{\dfrac{9.08 \times 10^7}{1.76 \times 10^4}}$

$= 71.8 \text{ mm}$

(iv) $r_{YY} = \sqrt{\dfrac{I_{YY}}{A}}$

$= \sqrt{\dfrac{4.72 \times 10^7}{1.76 \times 10^4}}$

$= 51.8 \text{ mm}$

(v) $Z_{XX} = \dfrac{I_{XX}}{y}$

where y in this case is the distance from XX to the base of the T.
That is $y = 240 - 74.5$
$= 165.5$

Thus $Z_{XX} = \dfrac{7.72 \times 10^7}{165.5}$

$= 5.49 \times 10^5 \text{ mm}^3$

Worked example 9.4
Calculate I_{XX}, I_{YY}, r_{XX} and r_{YY} for the I section shown in Fig. 9.14. What is the section modulus Z_{XX} for the point P?

Solution
Since this section is symmetrical, the centroid G lies at the intersection of the two axes of symmetry.
I_{XX} may be found by two methods.

Method 1
Divide the section into three rectangles. For rectangles B and D

$I_{XX} = I_{CG} + Ac^2$

where $I_{CG} = \dfrac{bd^3}{12} = \dfrac{70 \times 9^3}{12} = 4252 \text{ mm}^4$

and $c = 90 - 4.5 = 85.5 \text{ mm}$

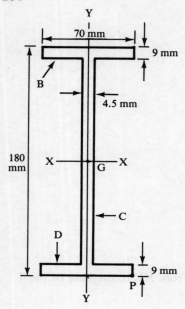

Fig. 9.14 Worked example 9.4

That is, for B and D, $I_{xx} = 4252 + 70 \times 9 \times 85.5^2$
$$= 4.61 \times 10^6 \text{ mm}^4$$

For rectangle C

$$I_{xx} = \frac{bd^3}{12}$$

$$= \frac{4.5 \times (180-18)^3}{12}$$

$$= 1.59 \times 10^6 \text{ mm}^4$$

Total $I_{xx} = 2 \times 4.61 \times 10^6 + 1.59 \times 10^6$
$$= 10.8 \times 10^6 \text{ mm}^4$$

Method 2

I_{xx} for the I section is I_{xx} for the large rectangle minus I_{xx} for the two shaded rectangles – Fig. 9.15.

Each shaded rectangle is $(180 - 2 \times 9) = 162$ mm in depth by
$(70 - 4.5)/2$
$= 32.75$ mm in breadth.

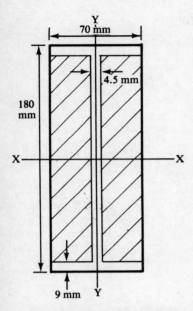

Fig. 9.15 Worked example 9.4

Thus $I_{XX} = \dfrac{70 \times 180^3}{12} - 2\left(\dfrac{32.75 \times 162^3}{12}\right)$

$= 3.40 \times 10^7 - 2.32 \times 10^7$
$= 10.8 \times 10^6 \text{ mm}^4$ as before.

For rectangles B and D

$I_{YY} = \dfrac{db^3}{12}$

$= \dfrac{9 \times 70^3}{12}$

$= 2.57 \times 10^5 \text{ mm}^4$

For rectangle C

$I_{YY} = \dfrac{db^3}{12}$

$= \dfrac{162 \times 4.5^3}{12}$

$= 1.23 \times 10^3 \text{ mm}^3$

Total I_{YY} = $2.57 \times 10^5 \times 2 + 1.23 \times 10^3$
 = 5.15×10^5 mm^4

$$r_{XX} = \sqrt{\frac{I_{XX}}{A}}$$

$$= \sqrt{\frac{10.8 \times 10^6}{2 \times 9 \times 70 + 162 \times 4.5}}$$

= 73.7 mm

$$r_{YY} = \sqrt{\frac{I_{YY}}{A}}$$

$$= \sqrt{\frac{5.15 \times 10^5}{2 \times 9 \times 70 + 162 \times 4.5}}$$

= 16.1 mm

For the point P, y = 90 mm

Thus section modulus for P is

$$Z_{XX} = \frac{I_{XX}}{y}$$

$$= \frac{10.8 \times 10^6}{90}$$

= 1.2×10^5 mm^3

Exercises

Objective type
Choose the ONE response which is the most appropriate.

9.1 The position of the neutral axis of a beam section depends on
    A the material of the beam
    B the area of the section
    C the shape of the section
    D the load on the beam

9.2 The second moment of area of a beam section
    A depends on the beam material
    B is a measure of the beam's resistance to bending
    C for a given area is a maximum for a square section
    D increases as the length of the beam increases

9.3 The second moment of area of a section could be measured in
  A mm^3
  B m^3
  C cm^2
  D mm^4

9.4 Section modulus may be defined as
  A I/y
  B $\sqrt{I/y}$
  C I/A
  D $\sqrt{I/A}$

9.5 For a rectangular beam section with depth d and breadth b, the section modulus at the upper surface is
  A $bd^2/12$
  B $bd^2/6$
  C $bd^3/12$
  D $bd^3/6$

9.6 Section modulus may be measured in
  A mm^3
  B cm^3
  C m^3
  D all of these

9.7 Radius of gyration r of a beam section may be defined as
  A $\sqrt{I/y}$
  B $\sqrt{A/I}$
  C $\sqrt{I/A}$
  D none of these

9.8 For a rectangular beam with depth d and breadth b, the radius of gyration r_{xx} of a section is
  A $d/6$
  B $d/12$
  C $d/\sqrt{3}$
  D $d/2\sqrt{3}$

9.9 The radius of gyration of a beam or column section
  A depends on the applied load
  B is one factor which determines the resistance to buckling
  C is larger for a steel beam than for a timber beam with the same section
  D all of these

Problems

9.10 Find the position of the centroid for the sections shown in Fig. 9.16 and Fig. 9.17.

9.11 Find the position of the principal axis XX for the section shown in Fig. 9.18.

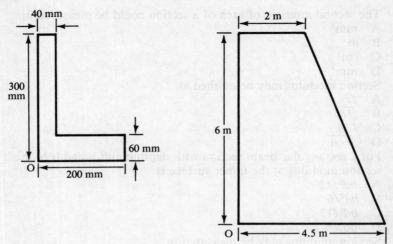

Fig. 9.16 Problem 9.10

Fig. 9.17 Problem 9.10

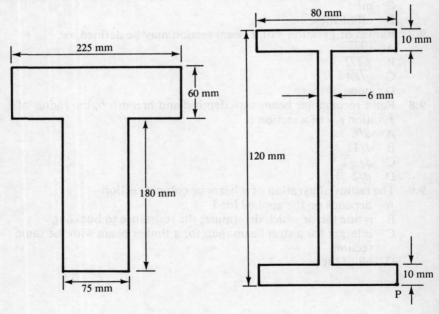

Fig. 9.18 Problem 9.11

Fig. 9.19 Problem 9.12

9.12 Find the second moment of area about the principal axes for the section in Fig. 9.19.

9.13 Find I_{XX} and I_{YY} for the section in Fig. 9.20.
9.14 Find the position of the centroid and hence the second moments of area about the major axes for the section in Fig. 9.21.

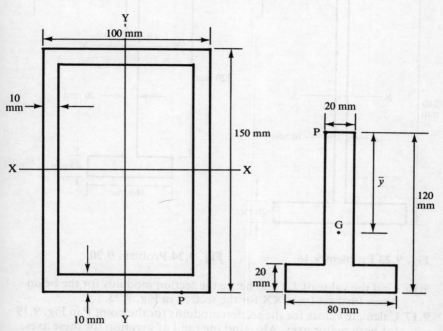

Fig. 9.20 Problem 9.13

Fig. 9.21 Problem 9.14

9.15 Find the value of I_{XX} for the section in Fig. 9.22(a). If the three parts of the section were rearranged as in Fig. 9.22(b), what would be the value of I_{XX}? Comment on the two answers obtained.

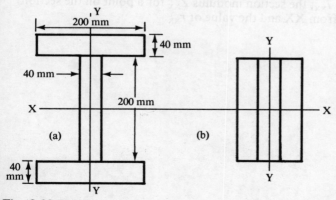

Fig. 9.22 Problem 9.15

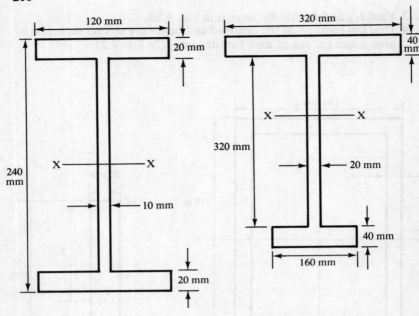

Fig. 9.23 Problem 9.16   **Fig. 9.24** Problem 9.20

9.16 Find the value of I_{xx} and hence the section modulus for the beam fibres furthest from XX for the section in Fig. 9.23.

9.17 Calculate values for the section modulus for the point P in Fig. 9.19 for both major axes. Also find the radii of gyration for these axes.

9.18 Find values for Z_{xx} and for Z_{yy} for the point P in the section in Fig. 9.20. Also find r_{xx} and r_{yy}.

9.19 Find the section modulus for the point P with respect to the major axis XX of the section in Fig. 9.21 and the radii of gyration.

9.20 For the section shown in Fig. 9.24, find the position of the principal axis XX, I_{xx}, the section modulus Z_{xx} for a point on the section furthest from XX and the value of r_{xx}.

Chapter 10

Simply-supported beams

Learning objectives

After reading this chapter and working through the exercises you should be able to:
- calculate reaction, shear force and bending moment values for simply supported beams with, or without, overhanging ends subject to point loads or uniformly distributed loads and sketch shear force and bending moment diagrams;
- calculate and plot bending stress distributions for rectangular, I and T sections subject to concentric loading;
- use techniques already developed to determine either safe loadings or beam size requirements.

10.1 Bending moment

When a beam is loaded, the loads produce moments which tend to bend the beam. These moments are called bending moments. For example, consider a simply supported beam (that is, with no lateral forces at the supports) loaded at its mid-point as shown in Fig. 10.1.

By symmetry considerations, R_1 and R_2 are both $W/2$. (The self-weight of the beam is ignored). The bending moment at XX due to R_1 acting on the shaded beam portion to the left of XX will be $W/2$ times the distance of XX from R_1 in a clockwise sense. If the beam is in equilibrium, then there is also a moment acting at XX on the shaded

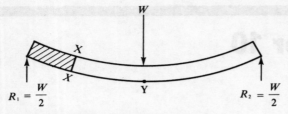

Fig. 10.1 Bending of a beam

portion set up by forces in the portion of the beam to the right of XX with the same magnitude as that due to R_1 but in an anticlockwise sense. The bending moment will be a maximum when XX is at the midpoint of the beam. If the stress produced by the bending is sufficiently large the beam material will fail at the point Y.

For any number of loads, the bending moment at any transverse beam section can be found using

Bending moment = algebraic sum (that is, taking sense into account) of the moments of all the forces acting on *either* side of the section

Bending moment may be measured in Nm, kNm, etc.

10.2 Bending moment sign convention

Since it is necessary to include the sense of a moment when calculating bending moment, a sign convention must be used. In this text the sign convention shown in Table 10.1 will be adopted.

Table 10.1—Bending moment sign convention

Sign convention	
+ (Positive)	− (Negative)
Beam upper surface concave upwards, that is, 'sagging' tendency	Beam upper surface convex upwards, that is, 'hogging' tendency

10.3 Shearing force

A beam may fail not only due to excessive bending but also by shearing across its length. Figure 10.2 shows a simple cantilever (that is, a beam supported at one end only) loaded at its free end.

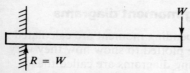

Fig. 10.2 Simple cantilever

To maintain equilibrium, reaction R at the fixed end must be W (ignoring again the self-weight of the beam). The load W must be transmitted through the beam material to the support. This sets up a force within the beam called the shearing force. If the beam material is not sufficiently strong then this shearing force causes the beam to fail as shown in Fig. 10.3.

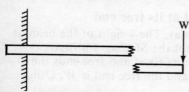

Fig. 10.3 Shearing of a beam

For any number of loads, the shearing force at any transverse beam section may be found using

Shearing force = algebraic sum (that is, taking direction into account) of all the forces acting on *either* side of the section

Shearing force may be measured in N, kN, etc.

10.4 Shearing force sign convention

Since the calculation of shearing force involves the direction of the loads and reactions, a sign convention for the various forces is necessary. The one adopted in this text is shown in Table 10.2.

Table 10.2—Shearing force sign convention

Sign convention	
+ (Positive)	− (Negative)
Up to left, down to right	Down to left, up to right

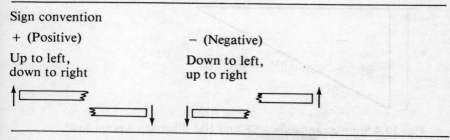

10.5 Shearing force and bending moment diagrams

When the values of shearing force and bending moment are calculated at points along the beam and graphs are plotted to show how they vary along the length of the beam, the resulting diagrams are called the bending moment (BM) and shearing force (SF) diagrams.

10.6 The four standard cases

There are four simple cases which can easily be analysed and which may be used as the bases for more complex situations. These will be dealt with separately.

(a) Cantilever with a single point load at its free end

The situation is as illustrated in Fig. 10.4(a). The weight of the beam is assumed to be negligible. It is usual to plot the SF diagram first.

At any point X on the beam between its fixed and free ends the shearing force on the section between X and the free end is W. Using

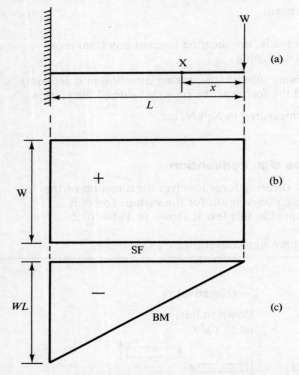

Fig. 10.4 Simple cantilever – SF and BM diagrams – point load

the sign convention of Table 10.2, the shearing force is positive and the SF diagram is thus as in Fig. 10.4(b).

At X, the bending moment is Wx and hence the bending moment increases linearly from zero at the free end to WL at the fixed end. Using the sign convention of Table 10.1, the BM diagram is as in Fig. 10.4(c).

(b) Cantilever with a uniformly distributed load

A load may be concentrated over a small area (point load) or spread uniformly along the length of a beam. In the latter case, the load is known as a uniformly distributed load (UDL).

In this case (Fig. 10.5) the UDL might be the self weight of the beam or it could be an imposed load or a combination of both. Let the UDL be w per unit length of beam. For the shaded section, shearing force will be wx. This means that shearing force varies linearly from 0 at the free end to wL at the fixed end. The SF diagram is as shown in Fig. 10.5(b).

To find the bending moment at X, the UDL on the shaded section is considered to act at its mid-point. So bending moment is $wx(x/2)$ or $wx^2/2$. This varies from zero at the free end to $wL^2/2$ at the fixed end but the variation is parabolic – Fig. 10.5(c).

The total load of W is wL and the maximum shearing force could thus be written as W and the maximum bending moment $wL^2/2$ as $WL/2$.

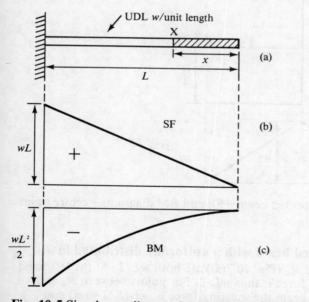

Fig. 10.5 Simple cantilever – SF and BM diagrams – uniformly distributed load

(c) Simply supported beam with a centre-point load

In this case (Fig. 10.6(a)), $R_A = R_B = W/2$. For the beam section to the left of W, the shearing force is just R_A or $W/2$ and is positive. For the section to the right of W, shearing force is $W/2$ and is negative – see Fig. 10.6(b).

The bending moment at both supports is zero. At any point X, distance x from the right hand (or left hand) support, the bending moment will be $Wx/2$. The maximum bending moment is thus $WL/4$ at the centre point where $x = L/2$. Since the beam is concave upwards, the bending moment is positive.

Note that when the bending moment is a maximum, the shearing force is zero.

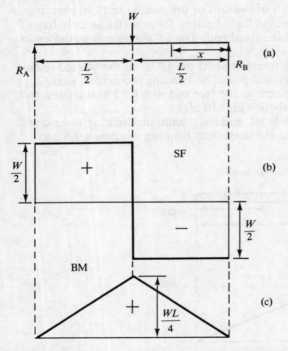

Fig. 10.6 Simply supported beam – SF and BM diagrams – centre point load

(d) Simply supported beam with a uniformly distributed load

The reactions R_A and R_B (Fig. 10.7(a)) are both $wL/2$. At the left hand support the shearing force is thus $wL/2$. For points between R_B and the mid-point of the beam the shearing force is $-(wL/2 - wx)$ and between R_A and the mid-point it is $+(wL/2 - wx)$, that is, it varies linearly with x. The SF diagram is thus as in Fig. 10.7(b).

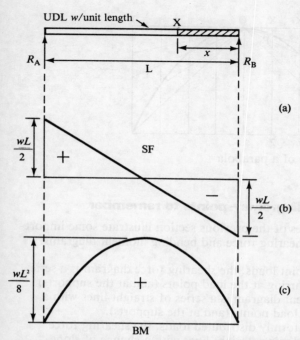

Fig. 10.7 Simply supported beam – SF and BM diagrams – uniformly distributed load

The bending moment at any point between the supports distance x from R_B is $wLx/2 - wx(x/2)$ or $wLx/2 - wx^2/2$. The variation of this with x is parabolic as in Fig. 10.7(c). The maximum value of the bending moment is when $x = L/2$ and is $wL^2/4 - wL^2/8$ or $wL^2/8$.

The maximum shearing force can also be written as $W/2$ and the maximum bending moment as $WL/8$ where W is the total load wL.

Note that when the bending moment is a maximum, the shearing force is zero.

10.7 Drawing a parabola

A simple geometrical method for drawing a parabolic curve is illustrated in Fig. 10.8 for drawing the parabola in Fig. 10.7(c).

The line XZ representing the maximum bending moment is divided into a number of equal parts. The line XO representing half the beam width is divided into the same number of parts. The radial lines OA, OB, etc, are then drawn together with the verticals X_1Z_1, X_2Z_2, etc. The intersection points Z, P_1, P_2, etc are points on the parabola which can then be drawn.

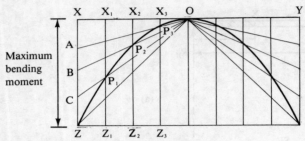

Fig. 10.8 Construction of a parabola

10.8 SF and BM diagrams – points to remember

The four standard cases of the previous section illustrate some important facts concerning shearing force and bending moment diagrams. These are as follows.

For a system of point loads, the shearing force diagram is a series of steps, the steps occurring at the load points (and at the supports) and the bending moment diagram is a series of straight lines with a change of slope at the load points (and at the supports).

For a series of uniformly distributed loads, the shearing force diagram is a series of sloping straight lines with a change of slope where the uniformly distributed load changes and the bending moment diagram is a curve. A single uniformly distributed load gives a parabolic bending moment curve.

These facts should be borne in mind when dealing with complex loading systems with both point and uniformly distributed loads.

Worked example 10.1

Deduce values for and hence draw the shearing force and bending moment diagrams for the cantilever loaded as in Fig. 10.9(a).

Solution

(a) Shearing force
For points between A and B, the shearing force is (30 + 20 + 10) = 60 kN down to the right, that is, 60 kN positive. For points between B and C, the shearing force is (20 + 10) = 30 kN positive and between C and D it is 10 kN positive. The SF diagram is thus as shown in Fig. 10.9(b).

(b) Bending moment
Since the loads are point loads, the BM diagram will be linear between loads. Also, all values of bending moment will be negative since the cantilever is convex upwards.

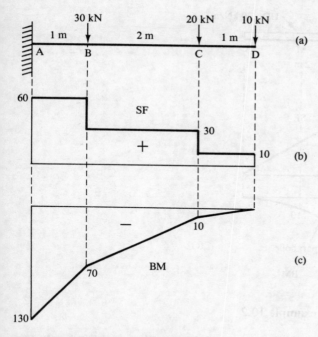

Fig. 10.9 Worked example 10.1

At A, bending moment = $-(30 \times 1 + 20 \times 3 + 10 \times 4) = 130$ kNm
At B, bending moment = $-(20 \times 2 + 10 \times 3) = -70$ kNm
At C, bending moment = $-(10 \times 1) = -10$ kNm
At D, bending moment = 0

The BM diagram is thus as shown in Fig. 10.9(c).

Worked example 10.2
Draw the shearing force and bending moment diagram for the cantilever loaded as in Fig. 10.10(a).

Solution
(a) Shearing force
For points between A and B, the shearing force $20 \times 2 = 40$ kN positive. At C, the shearing force is zero. Between B and C, the shearing force falls linearly (since the load is a UDL) from 40 kN to zero. The SF diagram is thus as in Fig. 10.10(b).

(b) Bending moment
All bending moments will be negative since the loading will cause the

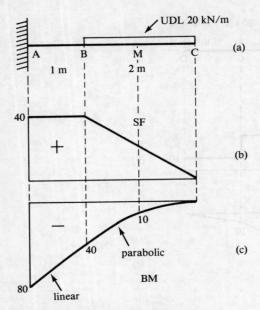

Fig. 10.10 Worked example 10.2

cantilever to be convex upwards. The UDL may be considered to be concentrated at M, the mid-point of BC.

At A, the bending moment is $-(20 \times 2 \times 2) = -80$ kNm
At B, the bending moment is $-(20 \times 2 \times 1) = -40$ kNm
At M, half-way between B and C, the bending moment is $-(20 \times 1 \times 0.5) = -10$ kNm.

Between A and B, the bending moment changes linearly from -80 kNm to -40 kNm. At C the bending moment is zero.

Between B and C, the variation of bending moment from -40 kNm to zero is parabolic. The BM diagram is thus as in Fig. 10.10(c).

Worked example 10.3
Calculate the reactions and draw the shearing force and bending moment diagrams for the simply supported beam shown in Fig. 10.11(a).

Solution
(a) Reactions
Take moments about A.
$5R_B = 10 \times 1.5 + 20 \times 3 + 15 \times 4$
$ = 135$

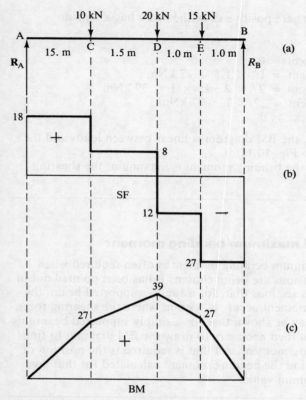

Fig. 10.11 Worked example 10.3

Thus $R_B = 27$ kN
Take moments about B.

$5R_A = 15 \times 1 + 20 \times 2 + 10 \times 2.5$
$ = 90$

Thus $R_A = 18$ kN
Check: $R_A + R_B = 27 + 18 = 45$ kN
$ = 10 + 20 + 15$ kN
$ =$ total load

(b) Shearing force
For points between A and C, shearing force $= R_A = 18$ kN positive
For points between C and D, shearing force $= 18 - 10 = 8$ kN positive
For points between D and E, shearing force $= 27 - 15 = 12$ kN negative
For points between E and B, shearing force $= R_B = 27$ kN negative

The SF diagram is thus as shown in Fig. 10.11(b).

(c) Bending moment
All bending moments are positive since the beam has a sagging tendency.

At A, bending moment = 0
At C, bending moment = $18 \times 1.5 = 27$ kNm
At D, bending moment = $27 \times 2 - 15 \times 1 = 39$ kNm
At E, bending moment = $27 \times 1 = 27$ kNm
At B, bending moment = 0

For point loads, the BM diagram is linear between loads and the diagram is thus as in Fig. 10.11(c).

Note that when the bending moment is maximum, the shearing force is zero.

10.9 Position of maximum bending moment

In practice, the maximum bending moment is often required when suitable beam dimensions are being chosen. It has been pointed out in examples in previous sections that, for a simply supported beam, the maximum bending moment occurs at a point where the shearing force is zero. In fact, it may be shown that for a simply supported beam this is always so. It is not then necessary to draw the BM diagram to find the maximum bending moment. All that is required is the position of zero shearing force and the bending moment calculated for that position is then its maximum value.

Worked example 10.4
Calculate the reactions for the loaded beam in Fig. 10.12(a), draw the shearing force and bending moment diagrams and find the value of the maximum bending moment.

Solution
(a) Reactions
Take moments about A (UDL assumed to act at the mid-point of AB)

$6R_B = 15 \times 1 + 5 \times 2.5 + 15 \times 4.5 + 8 \times 6 \times 3$
$= 239$

Thus $R_B = 39.8$ kN
Take moments about B.

$6R_A = 15 \times 1.5 + 5 \times 3.5 + 15 \times 5 + 8 \times 6 \times 3$
$= 259$

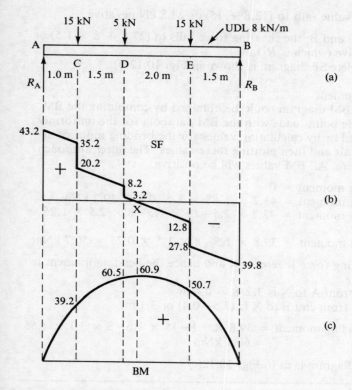

Fig. 10.12 Worked example 10.4

Thus R_A = 43.2 kN
Check: $R_A + R_B$ = 43.2 + 39.8 = 83.0 kN
                = total load (15 + 5 + 15 + 48)

(b) Shearing force
At A, shearing force = R_A = 43.2 kN positive.

Between A and C, the shearing force falls linearly due to the UDL at 8 kN/m and at C has fallen to (43.2 − 8) or 35.2 kN.

At C, the shearing force decreases abruptly by 15 kN to 20.2 kN.

Between C and D it falls linearly to (20.2 − 1.5 × 8) or 8.2 kN positive.

At D, there is a further abrupt decrease by 5 to 3.2 kN.

Between D and E, the shearing force becomes negative and the value falls to (3.2 − 2 × 8) or 12.8 kN negative.

At E, the value falls to (12.8 + 15) or 27.8 kN negative.

Between E and B, the shearing force falls to (27.8 + 8 × 1.5) or 39.8 kN negative (which is R_B).
The complete SF diagram is shown in Fig. 10.12(b).

(c) Bending moment
The complete BM diagram could be obtained by combining the BM diagram for the point loads with the BM parabola for the uniformly distributed load or by calculating values for the bending moment at suitable intervals and then plotting these values. The latter approach will be used here. All BM values will be positive.

At A, bending moment = 0
At C, bending moment = 43.2 × 1 − 8 × 1 × 0.5 = 39.2 kNm
At D, bending moment = 43.2 × 2.5 − 15 × 1.5 − 8 × 2.5 × 1.25 = 60.5 kNm
At E, bending moment = 39.8 × 1.5 − 8 × 1.5 × 0.75 = 50.7 kNm

The shearing force is zero at X and hence the bending moment is maximum here.
Distance from A to X is 3.2/8 = 0.4 m
Thus distance from end B to X is (3.5 − 0.4) or 3.1 m.

Maximum bending moment = 39.8 × 3.1 − 15 × 1.6 − 8 × 3.1 × 1.55
= 60.9 kNm

The BM diagram is as in Fig. 10.12(c).

Worked example 10.5
Draw the shearing force diagram for the beam in Fig. 10.13(a), deduce where the shearing force is zero and hence find the maximum bending moment.

Solution
(a) Reactions
Take moments about A.

$10R_B$ = 10 × 2 × 5 + 2 × 2 × 1 + 3 × 2 × 8.5 + 10 × 6
= 215

Thus R_B = 21.5 kN

Take moments about B.

$10R_A$ = 10 × 2 × 5 + 3 × 2 × 1.5 + 2 × 2 × 9 + 10 × 4
= 185

Thus R_A = 18.5 kN

Check: $R_A + R_B$ = 18.5 + 21.5 = 40 kN
= 2 × 2 + 10 × 2 + 3 × 2 + 10 = total load

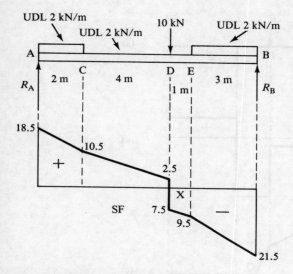

Fig. 10.13 Worked example 10.5

(b) Shearing force
At A, shearing force = R_A = 18.5 kN positive.

For points between A and C, the shearing force falls linearly at 4 kN/m (two UDLs of 2 kN/m each) from 18.5 kN positive at A to 10.5 kN positive at C.

Between C and D, the shearing force falls linearly at 2 kN/m to 2.5 kN positive at D.

At D, there is a drop of 10 kN to 7.5 kN negative.
Between D and E, there is a linear drop of 2 kN/m to 9.5 kN negative at E.
Between E and B, the shearing force again falls linearly at 4 kN/m from 9.5 kN negative at E to 21.5 kN negative at B (which is R_B).
The complete SF diagram is shown in Fig. 10.13(b).
The shearing force is zero at X, which is at the 10 kN point load.
At X, considering the right section of beam, bending moment

= 21.5 × 4 − 3 × 2 × 2.5 − 4 × 2 × 2
= 55 kNm

The maximum bending moment is thus 55 kNm.

10.10 Simply supported beam with overhanging ends

An example of this is shown in Fig. 10.14(a) for point loading.

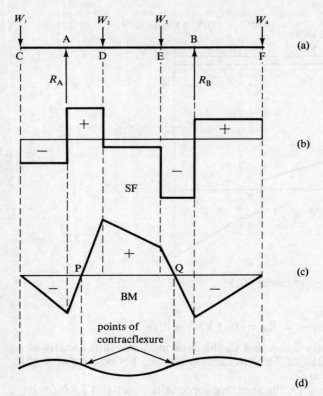

Fig. 10.14 Simply supported beam with overhang

The SF and BM diagrams might be as shown in Fig. 10.14(b) and (c). From Fig. 10.14(b), it is clear that the shearing force is zero at three points—the two supports and a point between the supports. At all three of these points the bending will pass through a maximum as shown in Fig. 10.14(c). The *absolute* maximum bending moment may occur at either of these three points depending on the loading.

At the points P and Q, the bending moment is zero. At these points, the bending is changing from a hogging tendency to a sagging tendency or vice versa. These points are known as points of contraflexure. The shape into which the beam would tend to bend is shown in Fig. 10.14(d).

Worked example 10.6
Find the reactions for the beam loaded as shown in Fig. 10.15(a) and draw the shearing force and bending moment diagrams. What is the maximum bending moment?

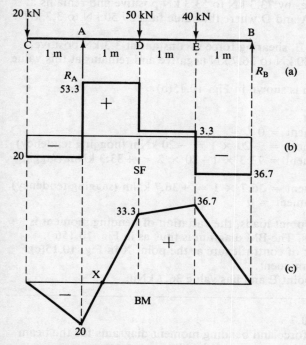

Fig. 10.15 Worked example 10.6

Solution
(a) Reactions.
Take moments about A.

$3R_B = 50 \times 1 + 40 \times 2 - 20 \times 1$
$= 110$

Thus $R_B = 36.7$ kN

Take moments about B.

$3R_A = 40 \times 1 + 50 \times 2 + 20 \times 4$
$= 220$

Thus $R_A = 73.3$ kN

Check : $R_A + R_B = 73.3 + 36.7 = 110$ kN
$= 20 + 50 + 40$
$=$ total load

(b) Shearing force
Between C and A, the shearing force is 20 kN negative.

At A, it increases by 73.3 kN to 53.3 kN positive and remains at this value between A and D where the value falls by 50 kN to 3.3 kN positive.

Between D and E, shearing force is constant at 3.3 kN positive, but at E it falls by 40 kN to 36.7 kN negative and remains at this value between E and B.

The SF diagram is shown in Fig. 10.15(b).

(c) Bending moment.
At C, bending moment = 0
At A, bending moment = $-20 \times 1 = -20$ kNm (hogging tendency)
At D, bending moment = $73.3 \times 1 - 20 \times 2 = +33.3$ kNm (sagging tendency)
At E, bending moment = $36.7 \times 1 = +36.7$ kNm (sagging tendency)
At B, bending momment = 0

Since loads are point loads, the variation of bending moment is linear between loads. The BM diagram is thus as in Fig. 10.15(c).

There is a point of contraflexure at the point X in Fig. 10.15(c).
Maximum bending moment.
This occurs at the point E and has value 36.7 kNm.

Worked example 10.7
Draw the shearing force and bending moment diagrams for the beam in Fig. 10.16(a). Find the maximum bending moment.

Solution
(a) Reactions
Take moments about A.

$4R_B = 6 \times 20 \times 3 - 1 \times 20 \times 0.5$
$= 350$

Thus $R_B = 87.5$ kN

Take moments about B.

$4R_A = 5 \times 20 \times 2.5 - 2 \times 20 \times 1$
$= 210$

Thus $R_A = 52.5$ kN

Check: $R_A + R_B = 52.5 + 87.5 = 140$ kN
$= 7 \times 20 =$ total load

(b) Shearing force
At C, shearing force = 0
Between C and A, the shearing force decreases linearly at a rate of 20 kN/m to a value of 20 kN negative at C (down to left).

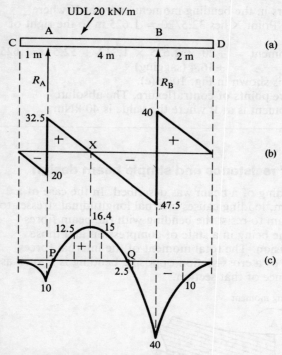

Fig. 10.16 Worked example 10.7

At A, the value increases by 52.5 kN to 32.5 kN positive.
Between A and B, the shearing force decreases at 20 kN/m to $(32.5 - 4 \times 20) = 47.5$ kN negative at B. Here, the value increases by 87.5 kN to 40 kN positive before decreasing at 20 kN/m to zero at D.
The shearing force diagram is shown in Fig. 10.16(b).

(c) Bending moment
The bending moment will be calculated at intervals of 1 m from C.

Distance from C in m	Bending moment in kNm
0 (at C)	0
1 (at A)	$-20 \times 1 \times 0.5 = -10$ (hogging)
2	$-20 \times 2 \times 1 + 52.5 \times 1 = 12.5$ (sagging)
3	$-20 \times 3 \times 1.5 + 52.5 \times 2 = +15$ (sagging)
4	$-20 \times 3 \times 1.5 + 87.5 \times 1 = -2.5$ (hogging)
5 (at B)	$-20 \times 2 \times 1 = -40$ (hogging)
6	$-10 \times 1 \times 0.5 = -10$ (hogging)
7 (at D)	0

A maximum occurs in the bending moment curve at X where shearing force is zero. Point X lies 32.5/20 = 1.625 m to the right of A.

At X, bending moment = $-20 \times 2.625 \times 1.312 + 52.5 \times 1.625$
= +16.4 (sagging)

The BM diagram is shown in Fig. 10.16(c).

Points P and Q are points of contraflexure. The absolute maximum bending moment is at B where the value is 40 kNm.

10.11 Moment of resistance and simple beam design

In Chapter 9, the bending of a beam was described. In the case of a simply supported beam, loading causes internal longitudinal stresses to develop within the beam to resist the bending with the beam fibres above the neutral plane being in a state of compression and those below in a state of tension. The total moment of the internal forces about the neutral axis of a cross-section opposing bending is known as the moment of resistance of that section.

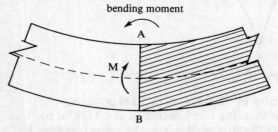

Fig. 10.17 Bending moment and moment of resistance

In Fig. 10.17, the bending moment and moment of resistance M are shown acting on the shaded portion of a beam. For equilibrium of the shaded section, bending moment and M are equal but opposite in sense.

Moment of resistance will now be considered further for a rectangular section beam.

Figure 10.18 shows the stress variation across a section AB.

It will be assumed that the Young's modulus for the beam material is the same in tension and in compression. The magnitude AA′ of the maximum compressive stress will then equal the magnitude BB′ of the the maximum tensile stress.

The resultant compressive force C will be the average compressive stress times the area which is $bd/2$ where the beam has dimensions b by d. If the maximum compressive (or tensile) stress is f then C is given by

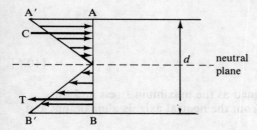

Fig. 10.18 Stress variation across a rectangular section

C = average stress × area

$$= \frac{f}{2} \times \frac{bd}{2}$$

that is, $C = \dfrac{fbd}{4}$ [10.1]

C will act at a distance $2/3\ (d/2)$ above the neutral plane.
Similarly, the resultant tensile force T is

$$T = \frac{fbd}{4} \qquad [10.2]$$

acting at a distance $2/3\ (d/2)$ below the neutral plane.

Now C and T are equal and opposite forces and hence constitute a couple.

This couple has a moment which is the moment of resistance M.

Thus Moment of resistance = force × separation

$$= \frac{fbd}{4} \times \frac{2d}{3}$$

or moment of resistance $M = \dfrac{fbd^2}{6}$ [10.3]

For a beam in equilibrium, the bending moment equals the moment of resistance.

Further, for a rectangular section, the second moment of area I of the section about the neutral axis is

$$I = \frac{bd^3}{12} \qquad [10.4]$$

Thus $\dfrac{2I}{d} = \dfrac{bd^2}{6}$

and hence from [10.3]

$$M = f\left(\dfrac{2I}{d}\right) \qquad [10.5]$$

However, f has been defined as the maximum stress and this occurs when y, the distance from the neutral axis, is a maximum and this is when

$$y = \dfrac{d}{2} \qquad [10.6]$$

Hence from [10.5] and [10.6]

$$M = \dfrac{fI}{y}$$

or $\dfrac{M}{I} = \dfrac{f}{y}$ \qquad [10.7]

Although this formula was derived in a simple manner for a rectangular section, a rigorous analysis yields the same answer for any section.

Equation [10.7] applies only to situations where there are no shear stresses, that is, to pure (or simple) bending. However, it is widely used with sufficient accuracy for ordinary bending problems involving both bending stresses and shear stresses. The formula may be written as

$$M = fZ \qquad [10.8]$$

since $Z = \dfrac{I}{y}$ the section modulus.

This is the basic formula for beam design. To summarize, M is the moment of resistance of a section, which for a beam in equilibrium equals the bending moment at that section (measured, for example, in Nmm), f is the maximum allowable stress (tensile or compressive, measured for example in N/mm^2) and Z is the section modulus for the fibres furthest from the neutral axis (measured, for example, in mm^3).

Worked example 10.8
A timber beam has section 50 mm wide by 175 mm deep. Its effective span is 3 m. The maximum permissible stress is 6 N/mm^2. Find the maximum safe centre point load (ignore the self weight of the beam).

Solution
$M = fZ$

$$Z = \frac{bd^2}{6} = \frac{50 \times 175^2}{6}$$

$$= 2.55 \times 10^5 \text{ mm}^3$$

Thus $M = 6 \times 2.55 \times 10^5$
$= 1.53 \times 10^6$ Nmm

Now the maximum bending moment for centre point loading is

$$\text{BM} = \frac{WL}{4}$$

$$= \frac{W \times 3000}{4} \quad (L \text{ in mm})$$

where W is the load.

Thus $\dfrac{3000W}{4} = 1.53 \times 10^6$

and $W = 2.04 \times 10^3$ N or 2.04 kN

The maximum safe centre-point load is thus 2.04 kN.

Worked example 10.9
A beam with section as in Fig. 10.19 carries a uniformly distributed load of 60 kN over a span of 6 m. It also has a central point load of 20 kN. Find the maximum stress due to bending.

Solution
For the section

$$I_{\text{XX}} = \frac{180 \times 250^3}{12} - 2\left(\frac{85 \times 210^3}{12}\right)$$

$$= 2.343 \times 10^8 - 1.312 \times 10^8$$
$$= 1.03 \times 10^8 \text{ mm}^4$$

where $y = 250/2 = 125$ mm

$$Z_{\text{XX}} = \frac{I_{\text{XX}}}{y}$$

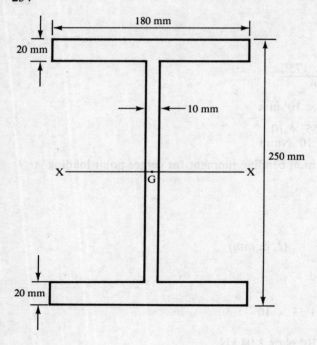

Fig. 10.19 Worked example 10.9

Thus $Z_{XX} = \dfrac{1.03 \times 10^8}{125}$

$= 8.24 \times 10^5 \text{ mm}^3$

Maximum bending moment due to the UDL $= \dfrac{WL}{8}$

$= \dfrac{60 \times 6}{8}$

$= 45 \text{ kNm}$

Maximum bending moment due to the central point load $= \dfrac{WL}{4}$

$= \dfrac{20 \times 6}{4}$

$= 30 \text{ kNm}$

Maximum bending moment $M = 30 + 45$
$= 75$ kNm
$= 75 \times 10^3 \times 10^3$
$= 75 \times 10^6$ Nmm

$M = fZ_{xx}$

Thus $75 \times 10^6 = f \times 8.24 \times 10^5$

Thus $f = \dfrac{75 \times 10^6}{8.24 \times 10^5}$

$= 91.0$ N/mm²

The maximum bending stress is thus 91.0 N/mm²

Worked example 10.10
Find the required section modulus and suggest suitable dimensions for a rectangular section timber beam given the following:

Span = 3.6 m

Loading: 5 kN uniformly distributed and 2 kN point centre load

Working stress = 7 N/mm²

Solution

Maximum bending moment due to UDL $= \dfrac{WL}{8} = \dfrac{5 \times 3.6}{8} =$

2.25 kNm

Maximum bending moment due to point load $= \dfrac{WL}{4} = \dfrac{2 \times 3.6}{4} =$

1.8 kNm

Maximum bending moment $M = 2.25 + 1.8$
$= 4.05$ kNm
$= 4.05 \times 10^3 \times 10^3$
$= 4.05 \times 10^6$ Nmm

$M = fZ$

Required $Z = \dfrac{4.05 \times 10^6}{7}$

$= 5.79 \times 10^5$ mm³

Section modulus required is thus 5.79×10^5 mm³. Now for a rectangular section

$$Z = \frac{bd^2}{6}$$

Thus $bd^2 = 6Z = 3.47 \times 10^6$ mm^3

The values of b and d must now be chosen so that $bd^2 = 3.47 \times 10^6$ mm^3.

If $b = 100$ mm then

$$d = \left(\frac{3.47 \times 10^6}{100}\right)^{\frac{1}{2}}$$

$= 186$ mm

If $b = 75$ mm then

$$d = \left(\frac{3.47 \times 10^6}{75}\right)^{\frac{1}{2}}$$

$= 215$ mm

Further, $b = 50$ mm gives $d = 263$ mm

In practice, the ratio of d to b lies between about 2:1 and 3:1. The values $d = 186$ mm, $b = 100$ mm and $d = 215$ mm, $b = 75$ mm would both be suitable. The standard sections 200 mm × 100 mm or 225 mm × 75 mm could be used.

Worked example 10.11
A floor is carried by timber joints with an effective span of 4 m spaced at 400 mm centres. The floor load is 5 kN/m². If the maximum permissible bending stress is 7.5 N/mm², find the minimum joist size – assume that the depth is to be three times the breadth.

Solution

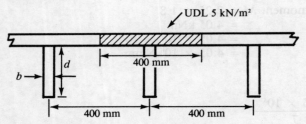

Fig. 10.20 Worked example 10.11

Each metre length of joist carries a load of $(1 \times 0.4) \times 5 = 2$ kN, that is, the UDL for the joists is 2 kN/m.

Maximum bending moment $M = \dfrac{WL}{8} = \dfrac{(2 \times 4) \times 4}{8}$

$\qquad\qquad\qquad\qquad\quad = 4$ kNm
$\qquad\qquad\qquad\qquad\quad = 4 \times 10^6$ Nmm

But $M = fZ$

Thus required $Z = \dfrac{4 \times 10^6}{7.5}$

$\qquad\qquad\qquad = 5.33 \times 10^5$ mm^3

Since $Z = \dfrac{bd^2}{6}$ and $d = 3b$

$5.33 \times 10^5 = \dfrac{9b^3}{6}$

Thus $b = 71$ mm
and hence $d = 213$ mm

The minimum joint size would thus be 71 mm × 213 mm. A beam with standard section 75 mm × 225 mm could be used.

Exercises

Objective type

Choose the ONE response which is the most appropriate.

10.1 Bending moment could be measured in
    A Nmm
    B kNm
    C MNm
    D all of these

10.2 For a simple cantilever, length L with a load W at its free end, the maximum bending moment is
    A $WL/2$
    B WL
    C WL^2
    D $WL^2/2$

10.3 For a simple cantilever, length L, with a total uniformly distributed load W, the maximum bending moment is
    A $WL/2$
    B WL
    C $WL^2/2$
    D WL^2

10.4 A simply-supported beam of length L has a point load W at a distance $L/4$ from one support. The maximum bending moment is
  A $WL/8$
  B $3WL/16$
  C $WL/4$
  D $WL/3$

10.5 A simply supported beam of length L has a uniformly distributed load of w/unit length. The maximum bending moment is
  A $wL^2/8$
  B $wL^2/4$
  C $wL/8$
  D $wL/4$

10.6 For the beam shown in Fig. 10.21 which statement is *not* true?

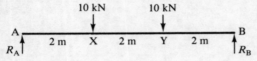

Fig. 10.21 Exercise 10.6

  A R_A and R_B are both 10 kN
  B The shearing force 1 m from A is 10 kNm
  C The bending moment at Y is 20 kNm
  D The shearing force half-way between X and Y is 20 kN

10.7 A point of contraflexure is a point where
  A the shearing force is zero
  B the shearing force is a maximum
  C the bending moment changes sign
  D the bending moment is a maximum

10.8 For a simply supported rectangular section beam, the moment of resistance M is
  A $fbd/6$
  B $f^2bd/6$
  C $fbd^2/6$
  D $fbd^2/12$

10.9 For simple bending
  A $Z = fM$
  B $My = fI$
  C $f = Mz$
  D $MI = fy$

10.10 The maximum bending moment for a beam is 50 kNm and the maximum allowable stress is 10 N/mm². The required section modulus is
  A 5×10^9 m³
  B 5×10^6 m³
  C 5×10^8 mm³
  D 5×10^6 mm³

Problems

10.11 to 10.13 For the cantilevers shown in Figs 10.22–10.24 draw shearing force and bending moment diagrams. State the value and position of the maximum bending moment in each case.

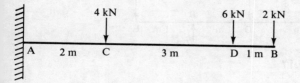

Fig. 10.22 Problem 10.11

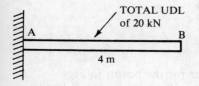

Fig. 10.23 Problem 10.12

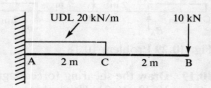

Fig. 10.24 Problem 10.13

10.14 to 10.16 For the beams shown in Figs 10.25–10.27 find the reactions at the supports and draw the shearing force and bending moment diagrams. Find the magnitude and position of the maximum bending moments.

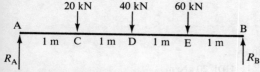

Fig. 10.25 Problem 10.14

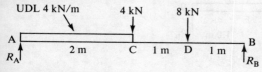

Fig. 10.26 Problem 10.15

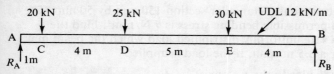

Fig. 10.27 Problem 10.16

10.17 to 10.18 Draw the shearing force and bending moment diagrams for the beams in Figs 10.28 and 10.29 and find the position of the points of contraflexure.

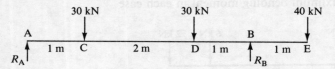

Fig. 10.28 Problem 10.17

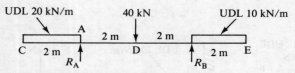

Fig. 10.29 Problem 10.18

10.19 to 10.20 Draw the shearing force diagrams for the beams in Figs 10.30–10.31. Using these, find the absolute maximum bending moment and state where it occurs in each case.

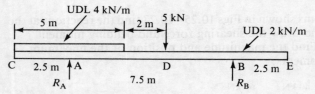

Fig. 10.30 Problem 10.19

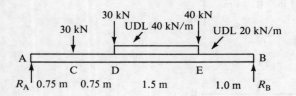

Fig. 10.31 Problem 10.20

10.21 A timber beam has rectangular section 150 mm by 50 mm. The maximum permissible bending stress is 7 N/mm². Find the maximum safe uniformly distributed load which the beam may carry over a 3.5 m span, if the load is applied

(a) perpendicular to the $x-x$ axis;
(b) perpendicular to the $y-y$ axis.

10.22 A cantilever 3 m long has a section as shown in Fig. 10.32. It carries a uniformly distributed load of 5 kN/m.
Find

(a) I_{xx} for the section;
(b) the maximum bending stress;
(c) the maximum additional point load at the free end if the maximum permissible bending stress is 100 N/mm².

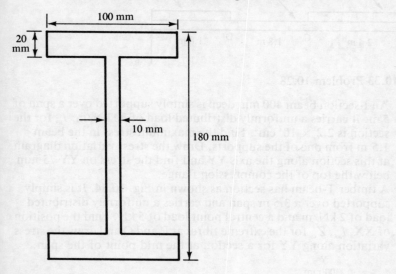

Fig. 10.32 Problem 10.22

10.23 Find the maximum safe point load for a beam spanning 3 m if the load is applied 1 m from a support and the maximum permissible stress is 6 N/mm². The beam section is 50 mm × 150 mm.

10.24 Find the required section modulus for a beam spanning 3.3 m if it is to carry a total uniformly distributed load of 4 kN. The working stress is 6.5 N/mm².

10.25 A timber beam is required to carry a point load of 10 kN at its midpoint. The span is 3.4 m. If the maximum permissible stress in the timber is 6 N/mm², find the required section modulus and suggest suitable dimensions for the beam section.

10.26 A timber cantilever is 2.8 m long and has a rectangular section 75 mm side by 200 mm deep. Find the total uniformly distributed load which the cantilever may carry if the bending stress is not to exceed 6.7 N/mm²

10.27 A floor has 75 mm × 200 mm timber joists spaced at 350 mm centres over a span of 3.5 m. The maximum permissible timber stress is 7 N/mm². The floor loading is 10 kN/m² uniformly distributed. Are the joists safe in bending?

10.28 Find the required section modulus for the simply supported beam in Fig. 10.33. The maximum permissible bending stress is 8 N/mm². Suggest a suitable beam depth if the width is to be 100 mm.

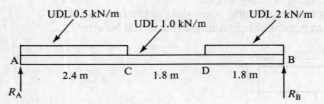

Fig. 10.33 Problem 10.28

10.29 An I-section beam 400 mm deep is simply supported over a span of 5 m. It carries a uniformly distributed load of 30 kN/m. I_{xx} for the section is 2.2×10^4 cm⁴. Find the maximum stress in the beam 1.5 m from one of the supports. Draw the stress variation diagram at this section along the axis YY and find the stress on YY 75 mm below the top of the compression flange.

10.30 A timber T-beam has section as shown in Fig. 10.34. It is simply supported over a 3.5 m span and carries a uniformly distributed load of 2 kN/m and a central point load of 5 kN. Find the position of XX, I_{xx}, Z_{xx} for the extreme fibres at P and Q and draw the stress variation along YY for a section at the mid-point of the span.

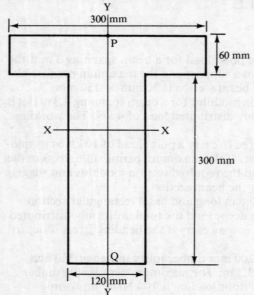

Fig. 10.34 Problem 10.30

Chapter 11

Soil retaining walls

Learning objectives

After reading this chapter and working through the exercises you should be able to:
- calculate pressures and forces on vertical retaining surfaces caused by granular soils;
- solve problems of spacing walings, props and struts in trench excavation supports.

In Chapter 5, the forces on surfaces retaining water were considered. In this chapter, the forces on surfaces retaining granular materials such as soil will be dealt with. Forces due to such materials are more difficult to find accurately than forces due to liquids. There are several reasons for this. The properties of soils, for example density, may not be uniform throughout the bulk of the retained material. Also, properties may change with time. For example, density may change due to change in moisture content. This change in density can lead to changes in other properties such as load bearing strength.

11.1 Pressure due to retained granular materials

There are several theories which may be used to find soil pressure on a retaining surface. Only one simple theory, the Rankine theory, will be used in this text. This theory applies only to granular materials like sandy soil, gravel, coke, etc. It is assumed that although there is internal friction between the particles of the material, there is no cohesion, that is, no force between the particles tending to bind them together. In a material such as a clay soil there is cohesion. It is further assumed that the materials are homogeneous and incompressible.

An important property of a granular material which is used in the Rankine theory is the angle of shearing resistance ϕ. This is a measure of the internal friction between the grains of the material. Values of ϕ for sand lie in the range 25°–40°. The value of ϕ for water is zero.

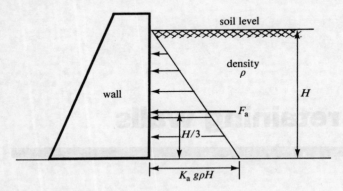

Fig. 11.1 Notation for Rankine theory

The Rankine theory predicts that the active lateral soil pressure p at a point on a vertical retaining surface at a depth h below the level surface of a soil with density ρ is

$$p_a = K_a \rho g h \qquad [11.1]$$

where K_a is called the coefficient of active earth pressure, given by

$$K_a = \frac{1 - \sin \phi}{1 + \sin \phi} \qquad [11.2]$$

At depth H,

$$p_a = K_a g \rho H$$

and hence the total active thrust F_a per metre run of wall (Fig. 11.1) is

$F_a = $ (average pressure) $\times H$

or $F_a = \tfrac{1}{2} K_a g \rho H^2 \qquad [11.3]$

This acts through the centroid of the pressure triangle, that is, $H/3$ above the base of the wall.

Equation [11.3] is sometimes written in the form

$$F_a = \tfrac{1}{2}K_a\gamma H^2 \qquad [11.4]$$

where γ is the specific weight.

Worked example 11.1
A wall with a vertical back 4 m high retains soil with density 2000 kg/m³ and an angle of shearing resistance of 30°. Find the force per unit run of wall due to the soil and the overturning moment due to this force about a point in the base.

Solution
Force per unit run $F_a = \tfrac{1}{2}K_a\, g\rho H^2$

$$K_a = \frac{1-\sin\phi}{1+\sin\phi}$$

$$= \frac{1-\tfrac{1}{2}}{1+\tfrac{1}{2}}$$

$$= \tfrac{1}{3}$$

Thus $F_a = \tfrac{1}{2} \times \tfrac{1}{3} \times 9.81 \times 2000 \times 4^2$
or $\quad F_a = 52.3$ kN/m

The force per unit run of wall is thus 52.3 kN acting at a height of 4/3 m above the base.
The overturning moment per metre run of wall is thus

$52.3 \times 4/3 = 69.7$ kNm

Worked example 11.2
Find the force on the brick wall shown in Fig. 11.2, due to the soil. The brick density is 2000 kg/m³, the specific weight of the soil is 18 kN/m³ and the angle of shearing resistance is 35°. Find also the

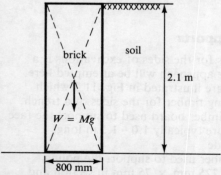

Fig. 11.2 Worked example 11.2

factor of safety against sliding if the coefficient of friction between the wall and its foundations is 0.7.

Solution
Force per unit run $F_a = \frac{1}{2} K_a \gamma H^2$

$$K_a = \frac{1-\sin\phi}{1+\sin\phi} = \frac{1-\sin 35°}{1+\sin 35°}$$

$$= 0.271$$

Thus $F_a = \frac{1}{2} \times 0.271 \times 18 \times 2.1^2$
or $F_a = 10.8$ kN/m

Force per unit run of wall is thus 10.8 kN/m.

Weight of unit run of wall is

$$W = Mg = 0.8 \times 2.1 \times 1 \times 2000 \times 9.81$$
$$= 3.30 \times 10^4 \text{ N/m}$$
$$= 33.0 \text{ kN/m}$$

Frictional force F between wall and foundations is

$F = \mu N$
$= \mu M g$
$= 0.7 \times 33.0$
$= 23.1$ kN/m run of wall

Since the horizontal force F_a tending to make the wall slide is 10.8 kN/m, the wall should be safe against sliding.

Factor of safety against sliding $= \dfrac{23.1}{10.8} = 2.1$

In practice the factor of safety against sliding should not be less than 1.5.

11.2 Trench excavation supports

The design of temporary supports for the sides of excavations is a complex matter and only a simple approach will be attempted here.

Some commonly used terms are illustrated in Fig. 11.3 which shows one method of support using timber for the sides of a trench.

A poling board is a vertical timber board used to support the face of an excavation. Poling boards are typically 1.0 – 1.5 m long, 25 – 50 mm thick and 225 mm wide.

A waling is a horizontal member used to support the poling boards. Typical walings could be 225 mm × 75 mm in section and

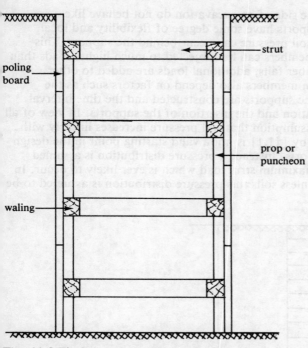

Fig. 11.3 Timber supports for an excavation

3.6 m long at 1–2 m centres. Sections larger than 300 mm × 300 mm are not normally used.

Walings are held against the poling boards by horizontal members known as struts. They are often of the same depth as the walings. A typical spacing for 225 mm × 75 mm struts could be at 1.5–2 m centres. A prop or puncheon is a vertical member (often placed against the poling boards) used to support a waling or strut on another one below it.

For excavations less than about 6 m deep, experience has shown that design of the timbering required is usually unnecessary. Choice of size is determined not only by experience but also by making sure that the sections chosen are sufficiently large to enable the members to be re-used several times and to withstand wear and tear.

11.3 Excavations more than 6 m deep

When excavations are more than 6 m deep, it may be economically desirable or necessary to calculate soil pressures. Only one method of calculating soil pressure will be considered here. This is one developed by Terzaghi and Peck. It applies only to cohesionless, dry soils.

Supports for the side of an excavation do not behave like a retaining wall. Supports have some degree of flexibility and local concentrations of soil pressure can occur behind the support. This means that some members can be subjected to much higher loads than others. If one member fails, additional loads are added to other members. Loads on members also depend on factors such as the manner in which the supports are constructed and the time interval between the excavation and the insertion of the supports. In view of all these factors the assumption that soil pressure increases linearly with depth as predicted by [11.1] is not a valid starting point in the design of excavation supports. Instead, a pressure distribution is assumed which covers the maximum strut load which is ever likely to occur. In the case of cohesionless soils this pressure distribution is assumed to be as in Fig. 11.4.

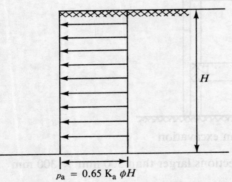

Fig. 11.4 Terzaghi and Peck rule for pressure distribution in cohesionless soils

The maximum pressure is given by the empirical formula

$$p_a = 0.65 K_a \gamma H \qquad [11.5]$$

This formula may be used to design suitable sizes and spacing for struts and walings. Struts may fail by crushing or, for long struts, by buckling. Walings can fail by being subjected to excessive bending stress.

A worked example will illustrate the use of [11.5].

Worked example 11.3
An excavation 7 m deep in cohesionless soil of specific weight 16 kN/m³ and with angle of shearing resistance 30° is supported by vertical timber poling boards, walings and struts. The walings to be used have section 300 mm × 150 mm and the struts have section 300 mm × 75 mm. Suggest suitable spacings for the walings and struts given the following data for the timber used.

Maximum bending stress = 6.0 N/mm²
Maximum compressive stress parallel to grain = 5.7 N/mm²
Maximum compressive stress perpendicular to grain = 2.2 N/mm²
Maximum shearing stress = 1.8 N/mm²

Solution

Maximum soil pressure is

$$p_a = 0.65\, K_a\, \gamma H$$

where $K_a = \dfrac{1 - \sin 30°}{1 + \sin 30°}$

$= \tfrac{1}{3}$

Thus $p_a = 0.65 \times \tfrac{1}{3} \times 16 \times 7$
$ = 24.3 \text{ kN/m}^2$

As a starting point, assume the walings spaced at 1 m centres and the struts at 1.5 m centres—see Fig. 11.5.

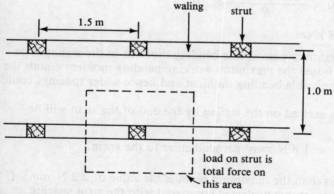

Fig. 11.5 Worked example 11.3

The following factors should be considered:

Compressive stress in the strut
Bending stress in the waling
Stress perpendicular to the grain in the waling
Shear stress in the waling

The maximum load of any strut is taken to be the total force on a rectangle with height equal to the waling spacing and width equal to the strut spacing. Thus

Maximum strut load = $24.3 \times 1.0 \times 1.5$
$ = 36.4$ kN

Thus the compressive stress in the strut

$$= \frac{36.4 \times 10^3}{300 \times 75}$$

$$= 1.6 \text{ N/mm}^2$$

This is well within the maximum permissible compressive stress of 5.7 N/mm².

The total uniformly distributed load on a waling between struts is 1 × 1.5 × 24.3 kN. If the waling is assumed to be simply supported between struts, then, for a waling,

$$\text{Maximum working bending moment} = \frac{WL}{8} = \frac{1.5 \times 24.3 \times 1.5}{8}$$

that is, $\text{BM}_{\text{MAX}} = 6.8$ kNm

Now $M = fZ$

Thus $M = 6.0 \times \dfrac{150^2 \times 300}{6} \times 10^{-6}$

$= 6.8$ kNm

This is the maximum permissible bending moment in the waling. For the spacings chosen the maximum working bending moment equals the maximum permissible bending moment and hence wider spacings could not be used.

The stress exerted on the waling by the end of the strut will be

$$\frac{36.4 \times 10^3}{300 \times 75} = 1.6 \text{ N/mm}^2 \text{ perpendicular to the grain.}$$

This is less than the maximum permissible value of 2.2 N/mm². If the width of the strut is neglected compared with the strut spacing, then the shearing force at a strut is half the strut load, that is, 18.2 kN. Thus

$$\text{Maximum shearing stress in waling} = \frac{18.2 \times 10^3}{150 \times 300}$$

$$= 0.4 \text{ N/mm}^2$$

This is well within the maximum shearing stress of 1.8 N/mm².
The initial choice of spacings of 1 m centres for the walings and 1.5 m centres for the struts are thus the greatest permissible and should be satisfactory.

This analysis neglects the fact that the struts would show a tendency to buckle. To allow for this, the maximum permissible compressive stress parallel to the grain should be reduced by a factor depending on the slenderness ratio of the strut. This treatment goes beyond the scope of this text.

Exercises

Objective type
Choose the ONE response which is the most appropriate.

11.1 A typical value for the angle of shearing resistance of a sandy soil could be
  A 10°
  B 30°
  C 60°
  D 90°

11.2 The specific weight of a soil with density 1500 kg/m³ in kN/m³, is
  A 14.7
  B 153
  C 1500
  D 14.7×10^3

11.3 If the angle of shearing resistance of a soil is 33° then the coefficient of active earth pressure is
  A 3.39
  B 0.54
  C 0.29
  D none of these

11.4 According to the Rankine theory, the total active thrust due to a depth H of soil per metre run of a vertical surface is
  A $K_a g\rho H$
  B $\tfrac{1}{2} K_a g\rho H$
  C $K_a g\rho H^2$
  D $\tfrac{1}{2} K_a g\rho H^2$

Problems

11.5 Soil of density 1600 kg/m³ and angle of shearing resistance 25° is retained by the vertical surface of a wall 4 m high. The soil is level with the top of the wall. Find the force on the wall, per m run, due to the soil.

11.6 A wall 3.6 m high, of rectangular section and density 2400 kg/m³ retains soil of specific weight 20 kN/m³ level with its top. If the angle of shearing resistance is 30°, find the thickness of wall required so that the factor of safety against overturning of the wall is 2.5.

11.7 A wall with a vertical back 5 m high retains sand to a depth of 4.8 m. The wall has trapezoidal section. The top width is 1 m and the base width is 3 m. If the soil density is 1800 kg/m³, the angle of shearing resistance is 35° and the wall density is 2300 kg/m³, find if the wall is safe against sliding when the coefficient of friction between it and its foundations is 0.55.

11.8 The sides of a trench 6 m deep in sand of specific weight 20 kN/m³ and angle of shearing resistance 35° are supported by poling boards, walings and struts. The walings available are 225 mm × 225 mm and the struts available are 200 mm × 150 mm. Walings are spaced at 1.5 m centres. The data appropriate to the timber used are as follows:

Maximum bending stress = 8.7 N/mm²
Maximum compressive stress parallel to grain = 6.9 N/mm²
Maximum compressive stress perpendicular to grain = 2.4 N/mm²
Maximum shearing stress = 1.8 N/mm²

Determine a suitable value for the strut spacing and state the working stresses.

11.9 An excavation 8 m deep in soil with density 1800 kg/m³ and angle of shearing resistance 30° is supported by timer poling boards, waling and struts with the same stress values as in Problem 11.8. The waling spacing is to be 1.3 m and the strut spacing 1.8 m. Suggest suitable timber sizes for the walings and struts.

Part 5

Answers

Objective type exercises

Chapter 1
1.1 D
1.2 A
1.3 B
1.4 C
1.5 D
1.6 A
1.7 B
1.8 D

Chapter 2
2.1 D
2.2 C
2.3 B
2.4 C
2.5 D
2.6 A
2.7 B
2.8 B
2.9 B
2.10 B
2.11 A
2.12 D
2.13 D

Chapter 3
3.1 C
3.2 B
3.3 B
3.4 D
3.5 D
3.6 D
3.7 A
3.8 A
3.9 C
3.10 B
3.11 C
3.12 C
3.13 C
3.14 B
3.15 C
3.16 D
3.17 B

Chapter 4
4.1 D
4.2 A
4.3 C
4.4 B
4.5 D
4.6 A
4.7 D

Chapter 5
5.1 D
5.2 B
5.3 C
5.4 D
5.5 D
5.6 B
5.7 C
5.8 B
5.9 B
5.10 A
5.11 C
5.12 D
5.13 C

Chapter 6
6.1 B
6.2 A
6.3 D
6.4 C
6.5 B
6.6 D
6.7 A
6.8 D
6.9 C
6.10 B
6.11 A
6.12 D
6.13 C
6.14 D
6.15 A

Chapter 7
7.1 B
7.2 C
7.3 A
7.4 B
7.5 C
7.6 C
7.7 D
7.8 C
7.9 D
7.10 A
7.11 B

Chapter 8
8.1 B
8.2 A
8.3 C
8.4 D
8.5 D
8.6 D

Chapter 9
9.1 C
9.2 B
9.3 D
9.4 A
9.5 B
9.6 D
9.7 C
9.8 D
9.9 B

Chapter 10
10.1 D
10.2 B
10.3 A
10.4 B
10.5 A
10.6 D
10.7 C
10.8 C
10.9 B
10.10 D

Chapter 11
11.1 B
11.2 A
11.3 C
11.4 D

Problems — Chapter 1

1.9   2/3 m/s
1.10   8.78 m/s^2
1.11   84.8 s
1.12   7.00 s
1.13   53.2 m
1.14   0.8 m/s^2; 9.55 kN
1.15   1 m/s; 101.1 s
1.16   200 s
1.17   20 s
1.18   12.57 rad/s; 6.28 m/s
1.19   464 m/s
1.20 -0.42 rad/s^2
1.21   104.7 rad/s^2; 31.4 m/s
1.22   0.463 m/s^2; 0.617 rad/s^2; 22.2 rad/s

Problems — Chapter 2

2.14   2.4 N
2.15   9.81 kN
2.16   2780 N
2.17   0.275 m/s^2
2.18   1.5 s; 1.12 m
2.19   687 N; 757 N; 617 N; 687 N
2.20   10 kN
2.21   50 kN
2.22   0.25 m/s^2; 0.185 m/s^2; 0.348 m/s^2
2.23   736 N
2.24   13.0 kN; 11.8 kN; 9.4 kN
2.25   250 kN; 0.662 m/s^2
2.26   1.09 m/s^2; 43.6 N
2.27   1.28 m/s^2; 22.2 N; 0.625 s
2.28   15.8°
2.29   350 mm
2.30   33.4 rev/min

Problems — Chapter 3

3.18   392 kJ
3.19   294 J
3.20   2200 kJ approx.
3.21   162 kJ approx.
3.22   9.81 kJ; -98.1 kJ
3.23   157 kJ
3.24   10.2 J
3.25   39.2 J

3.26	200 J; 200 J; 200 J
3.27	343 W
3.28	533 N
3.29	13.2 N; 6.6 m/s^2; 15.2 m; 149 J; 51 J
3.30	13.9 J; 8.175 kJ; 8.19 kW; 14.9 kW
3.31	58.3 m/s; 44.0 m/s
3.32	8.89 kN; 16.9 m/s
3.33	1360 MJ; 10.5 MJ; 63.4 kW
3.34	0.49
3.35	942 J
3.36	33.5 Nm
3.37	6.13 kJ; 487 Nm; 12.3 kW
3.38	911 N; 434 N
3.39	262 W
3.40	1.12 kW
3.41	2.97 × 10^4 J; 52.5 Nm

Problems — Chapter 4

4.8	100 N
4.9	43.7 kN
4.10	200 N
4.11	410 kN/m^2
4.12	180 kg; 32.4 kN
4.13	47.0 Ns; 1.93 kN
4.14	0.25 m
4.15	0.29 m/s; 2.29 m/s; both in the original directions
4.16	4.88 m/s; 0.010 s; 488 N; 0.295 J
4.17	8.44 m/s; 3.75 m/s
4.18	24.7 kJ

Problems — Chapter 5

5.14	51.6 m
5.15	441 kN/m^2
5.16	72.6 kN/m^2; 544 mm
5.17	787 mm
5.18	41.6 kN/m^2
5.19	157 kN, acting at a point 1.33 m below the water level
5.20	188 kN/m run
5.21	82.3 kN; 64.2 kNm
5.22	70.5 kN; 1.44 m above the bottom
5.23	30.8 kN; 4.01 m below the water surface
5.24	48.8 kN
5.25	2210 kN; 3.24 m above the bottom; 52 kN
5.26	117 kN; 60.1 kN
5.27	11.5 kNm
5.28	35.3 MN; 141 MNm; yes

Problems — Chapter 6

6.16 9.42×10^{-3} m^3/s; 9.42 litre/s; 9.42 kg/s
6.17 0.637 m/s
6.18 4.43 m/s; 1.97 m/s; 1.51 m/s; 4.10 m/s
6.19 30.6 kN/m^2
6.20 3.56 m/s; 335 kN/m^2
6.21 0.784 m
6.22 0.397 m
6.23 287 mm
6.24 0.771 m; 1.94 m/s; 9.43 m^3/s
6.25 1.22 m^3/s
6.26 3.23 m^3/s
6.27 2.5 m deep; 1:1760

Problems — Chapter 7

7.12 11.8 kN/m^2
7.13 54.4 kN/m^2
7.14 102.0 kN/m^2; 0.0036%
7.15 61.8 kN/m^2; 6.3 m
7.16 2.94 kN/m^2
7.17 392 N/m^2; 0.040
7.18 9.68 litre/s
7.19 0.155 m^3/s
7.20 0.91
7.21 1.58 m
7.22 7.31 litre/s
7.23 0.524 m
7.24 3.58 m^3/s
7.25 0.60

Problems—Chapter 8

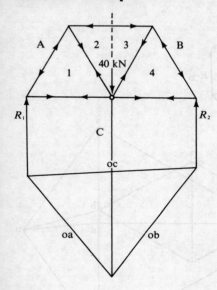
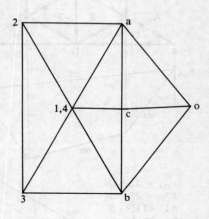

$R_1 = R_2 = 20$ kN   12   23.1 kN T
A1   23.1 kN S   34   23.1 kN T
A2/B3 23.1 kN S   C1   11.5 kN T
B4   23.1 kN S   C4   11.5 kN T

8.8

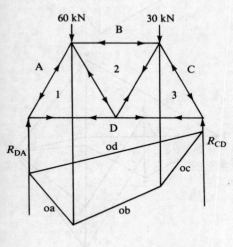
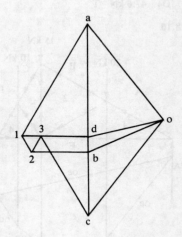

$R_{DA} = 52.5$ kN   12   8.7 kN   S
$R_{CD} = 37.5$ kN   23   8.7 kN   T
A1   60.6 kN S   D1   30.3 kN T
B2   26.0 kN S   D3   21.7 kN T
                          C3   43.3 kN S

8.9

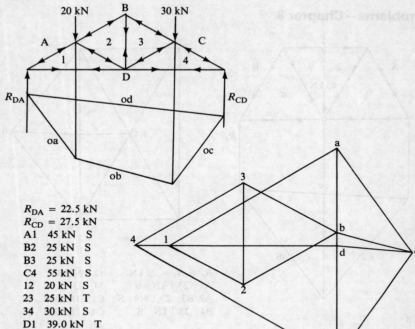

R_{DA} = 22.5 kN
R_{CD} = 27.5 kN
A1 45 kN S
B2 25 kN S
B3 25 kN S
C4 55 kN S
12 20 kN S
23 25 kN T
34 30 kN S
D1 39.0 kN T
D4 47.6 kN T

8.10

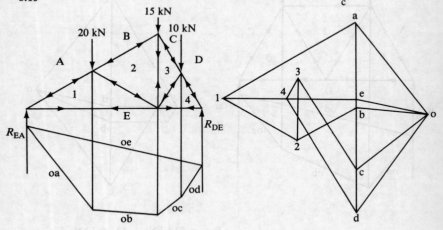

R_{EA} = 27.5 kN   C3 26.0 kN S   34 5.8 kN S
R_{DE} = 22.5 kN   D4 31.8 kN S   E1 30.3 kN T
A1 35.0 kN S   12 20.0 kN S   E4 15.9 kN T
B2 15.0 kN S   23 15.0 kN T

8.11

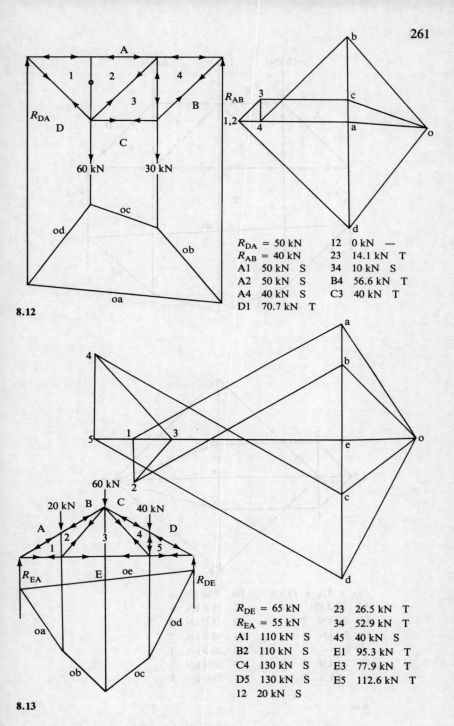

8.12

R_{DA} = 50 kN		12	0 kN —
R_{AB} = 40 kN		23	14.1 kN T
A1	50 kN S	34	10 kN S
A2	50 kN S	B4	56.6 kN T
A4	40 kN S	C3	40 kN T
D1	70.7 kN T		

8.13

R_{DE} = 65 kN		23	26.5 kN T
R_{EA} = 55 kN		34	52.9 kN T
A1	110 kN S	45	40 kN S
B2	110 kN S	E1	95.3 kN T
C4	130 kN S	E3	77.9 kN T
D5	130 kN S	E5	112.6 kN T
12	20 kN S		

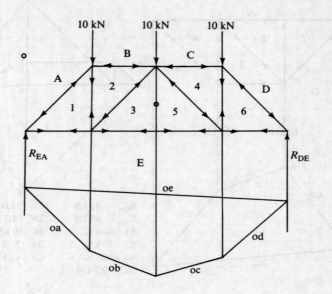

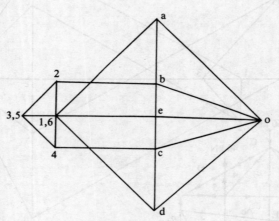

$R_{EA} = R_{DE} = 15$ kN	D6	21.2 kN	S
A1 21.2 kN S	B2	15.0 kN	S
12 5.0 kN T	C4	15.0 kN	S
23 7.1 kN S	E1	15.0 kN	T
35 0 —	E3	20.0 kN	T
45 7.1 kN S	E5	20.0 kN	T
46 5.0 kN T	E6	15.0 kN	T

8.14

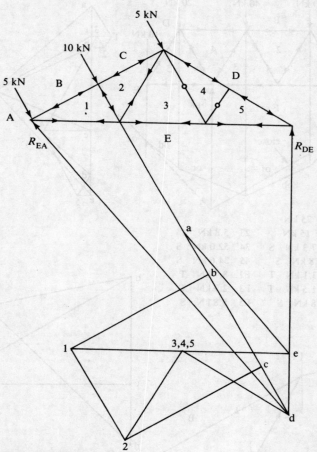

$R_{DE} = 5.8$ kN
$R_{EA} = 15.3$ kN 40° to horizontal
B1 14.4 kN S
C2 14.4 kN S
D4 11.5 kN S
D5 11.5 kN S
12 10 kN S
23 10 kN T
34 0 kN —
45 0 kN —
E1 20 kN T
E3 10 kN T
E5 10 kN T

8.15

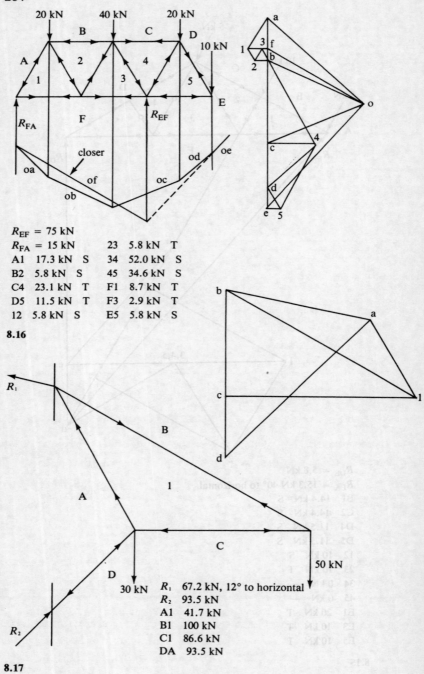

$R_{EF} = 75$ kN
$R_{FA} = 15$ kN

A1	17.3 kN S	23	5.8 kN T
B2	5.8 kN S	34	52.0 kN S
C4	23.1 kN T	45	34.6 kN S
D5	11.5 kN T	F1	8.7 kN T
12	5.8 kN S	F3	2.9 kN T
		E5	5.8 kN S

8.16

R_1 67.2 kN, 12° to horizontal
R_2 93.5 kN
A1 41.7 kN
B1 100 kN
C1 86.6 kN
DA 93.5 kN

8.17

265

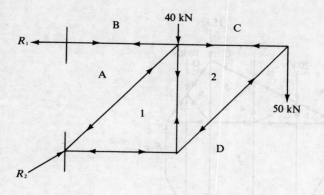

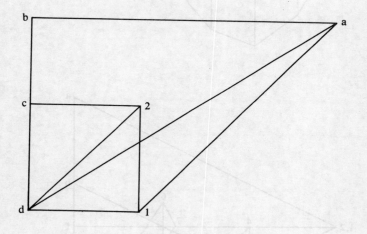

R_1 140.0 kN
R_2 166.2 kN, 33° to horizontal
AB   140.0 kN   T
C2   50.0 kN   T
D2   70.7 kN   S
D1   60.0 kN   S
A1   127.3 kN   S
12   50.0 kN   T

8.18

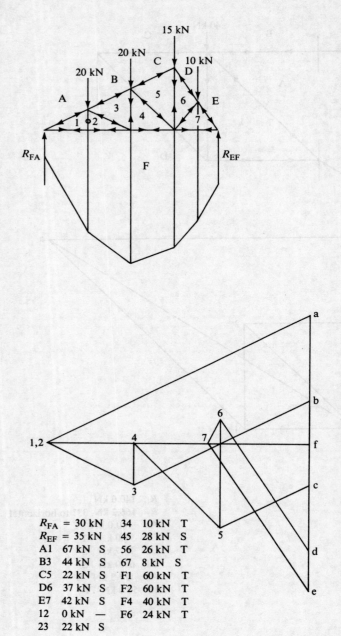

R_{FA} = 30 kN		34	10 kN T
R_{EF} = 35 kN		45	28 kN S
A1	67 kN S	56	26 kN T
B3	44 kN S	67	8 kN S
C5	22 kN S	F1	60 kN T
D6	37 kN S	F2	60 kN T
E7	42 kN S	F4	40 kN T
12	0 kN —	F6	24 kN T
23	22 kN S		

8.19

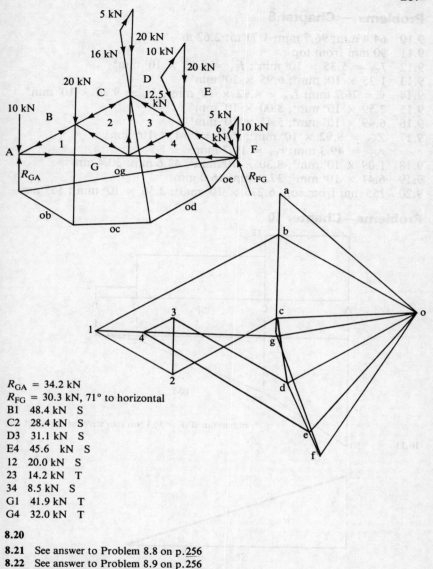

R_{GA} = 34.2 kN
R_{FG} = 30.3 kN, 71° to horizontal
B1 48.4 kN S
C2 28.4 kN S
D3 31.1 kN S
E4 45.6 kN S
12 20.0 kN S
23 14.2 kN T
34 8.5 kN   S
G1 41.9 kN T
G4 32.0 kN T

8.20

8.21 See answer to Problem 8.8 on p. 256
8.22 See answer to Problem 8.9 on p. 256
8.23 See answer to Problem 8.10 on p. 257
8.24 See answer to Problem 8.11 on p. 257
8.25 See answer to Problem 8.12 on p. 258
8.26 See answer to Problem 8.14 on p. 259
8.27 See answer to Problem 8.15 on p. 260
8.28 See answer to Problem 8.16 on p. 261
8.29 See answer to Problem 8.17 on p. 261
8.30 See answer to Problem 8.18 on p. 262

Problems – Chapter 9

9.10 64.4 mm; 96.7 mm; 1.71 m; 2.62 m
9.11 90 mm from top
9.12 $I_{XX} = 5.35 \times 10^6$ mm^4; $I_{YY} = 8.55 \times 10^5$ mm^4
9.13 1.35×10^7 mm^4; 6.95×10^6 mm^4
9.14 $\bar{y} = 76.7$ mm; $I_{XX} = 4.92 \times 10^6$ mm^4; $I_{YY} = 9.20 \times 10^5$ mm^4
9.15 2.59×10^8 mm^4; 8.00×10^7 mm^4
9.16 6.49×10^7 mm^4; 5.41×10^5 mm^3
9.17 $Z_{XX} = 8.92 \times 10^4$ mm^3; $Z_{YY} = 2.14 \times 10^4$ mm^3; $r_{XX} = 49.3$ mm; $r_{YY} = 19.7$ mm
9.18 1.09×10^5 mm^3; 8.30×10^4 mm^3; 42.6 mm; 30.4 mm
9.19 6.41×10^4 mm^3; 37.0 mm; 16.0 mm
9.20 155 mm from top, 6.275×10^8 mm^4; 2.56×10^6 mm^3; 157 mm

Problems – Chapter 10

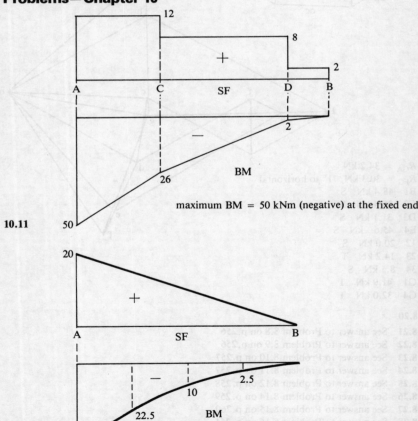

10.11 maximum BM = 50 kNm (negative) at the fixed end

10.12 maximum BM is 40 kNm (negative) at the fixed end

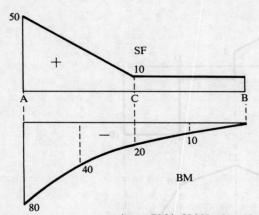

maximum BM is 80 kNm (negative) at the fixed end

10.13

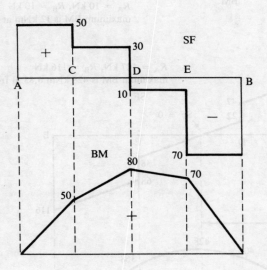

$R_A = 50$ kN, $R_B = 70$ kN
maximum BM is 80 kNm at the point D

10.14

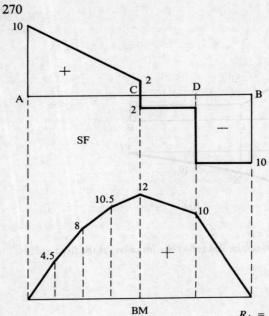

10.15

$R_A = 10$ kN, $R_B = 10$ kN
maximum BM is 12 kNm at the point C

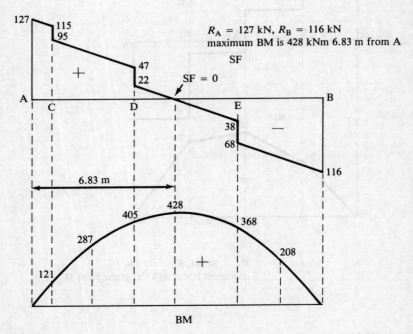

$R_A = 127$ kN, $R_B = 116$ kN
maximum BM is 428 kNm 6.83 m from A

10.16

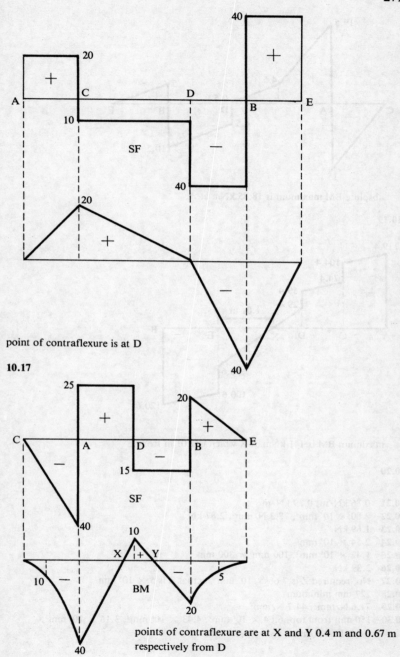

point of contraflexure is at D

10.17

points of contraflexure are at X and Y 0.4 m and 0.67 m respectively from D

10.18

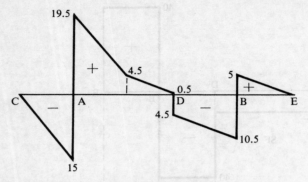

absolute BM maximum is 18.75 kNm at A

10.19

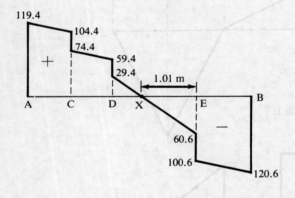

maximum BM is 141 kNm at X which is 1.01 m from E

10.20

10.21 0.86 kN/m, 0.29 kN/m
10.22 2.80×10^7 mm^4, 72.3 N/mm^2, 2.87 kN
10.23 1.69 kN
10.24 2.54×10^5 mm^3
10.25 1.42×10^6 mm^3, 100 mm × 300 mm
10.26 2.39 kN
10.27 No, required Z is 7.66×10^5 mm^3, actual Z is 5×10^5 mm^3
10.28 227 mm minimum
10.29 71.6 N/mm^2, 44.7 N/mm^2
10.30 150 mm from top. 6.64×10^8 mm^4, 4.43×10^6 mm^3, 3.16×10^6 mm^3, $f_P = 1.68$ N/mm^2, $f_Q = 2.35$ N/mm^2

Problems — Chapter 11

11.5	51.0 kN
11.6	1.75 m
11.7	Yes: μ required = 0.24
11.8	2.0 m; 8.3 N/mm^2; 2.1 N/mm^2; 2.1 N/mm^2; 0.6 N/mm^2
11.9	225 mm × 225 mm; 200 mm × 150 mm

Index

Acceleration 4
Angle of shearing resistance 244
Angular
  acceleration 10
  velocity 9

Barometer 87
Beam
  formula 232
  reactions 152
Belt drives 53
Bending moment
  definition 212
  diagrams 214
  maximum 222
  sign convention 212
  standard cases 214
Bernoulli equation 108
Bingham plastic 80
Bows notation 157
Bulk modulus 81

Cantilever frames 169
Capillarity 80
Centre of pressure 88, 89, 91
Centroid 88, 188
Channel flow 115
Chezy formula 115
Circular motion
  acceleration 12
  banked tracks 31
Coefficient of
  active earth pressure 244
  friction 29
  restitution 70
Components of a force 148
Compressibility 81
Connected masses 27
Conservation of
  energy 45
  linear momentum 67
Continuity equation 105
Contraflexure 226
Couple 48

Darcy-Weisbach formula 111
Density 78
Differential manometer 127
Discharge coefficient 133, 137, 139
Displacement 4
Distance-time curve 4

Earth pressure 243
Equilibrant 151
Equilibrium
  conditions for 152
  definition 151

Fluid
  definition 77
  pressure 82
Frameworks
  force diagram method 157
  pin-jointed 157
  semi-graphical method 164
Free fall 9
Friction
  coefficient 113
  factor 113
  laws of 29

Hydraulic
  gradient 116
  mean depth 116
Hydrostatic thrust 89, 91

Ideal fluid 80
Impact
  elastic 69
  inelastic 67
  of water on a fixed surface 65
  of water on a moving surface 66
Impulse of a force 62
Inclined
  loading 173, 179
  tube manometer 131
Inverted differential manometer 129

Joints, pin and rigid 153

Kinetic
 energy 43, 44, 71
 head 109

Laminar flow 104
Link polygon 159
Load
 concentrated 215
 uniformly distributed 215

Major axes 195
Mass 22
Mid-ordinate rule 40
Millibar 88
Moment of
 couple 48
 force 48, 152
 resistance 230
Momentum 23

Neutral
 axis 191
 plane 191
Newtonian fluid 79
Newton's laws of motion
 first 21
 second 22
 third 25
Newton's law of viscosity 79

Parabola construction 217
Parallel axis theorem 92, 198
Parallelogram rule 148
Piezometer 122
Pin-jointed frames 155
Plastic fluid 79
Poise-unit 79
Polar diagram 158
Poling board 246
Polygon
 of forces 153
 rule 149
Potential
 energy 42
 head 108
Power 46, 50, 53
Pressure
 absolute 86
 atmospheric 87
 definition 82
 gauge 86
 head 85, 109
 horizontal variation 83
 vertical variation 83
Principal axes 195
Prop 247
Properties of sections 188
Puncheon 247

Radius of gyration 196, 197
Rankine theory 243
Reynolds number
 channel flow 117
 pipe flow 111
Resolution of a force 148
Resultant 147
Retaining walls 243
Rolled steel joist 194

Scalar quantity 3
Second moment of area 90, 93, 194, 197
Section modulus 195, 197
Shearing force
 definition 213
 diagrams 214
 sign convention 213
 standard cases 214
Simpson's rule 41
Speed 4
Speed-time curve 5
Specific
 gravity 78
 weight 78
Static determinancy 154, 155
Steady flow 103
Stoke-unit 80
Strain 192
Streamlines 104
Stress 192
Struts 247
Surface tension 80
Superelevation 33

Tarzaghi and Peck rule 247
Thixotropic fluid 79
Torque 48
Trapezoidal rule 40
Triangle
 of forces 153
 rule 147
Turbulent flow 104

Uniform flow 103
Uniformly accelerated motion 6
U-tube manometer 123

Vapour pressure 82
Vector quantity 3
Velocity
 definition 4
 head 109
Venturi meter 132
Viscosity
 absolute 79
 dynamic 79
 kinematic 80

Waling 247
Weight 24
Weirs
  broad-crested 138
  sharp-crested 135, 137
  thin-plate 134
  V-notch 137

Work done by
  constant force 38
  constant torque 49
  variable force 39
  variable torque 51

Young's modulus 192